Century Farms of Williamson County

Century Farms of Williamson County

CHALLENGES, CHANGES, and CHOICES

CANETA SKELLEY HANKINS

Rick Warwick, Williamson County Historian
President, Williamson County Historical Society
Franklin, Tennessee

Century Farms of Williamson County Copyright © 2025
by Williamson County Historical Society
All Rights Reserved.

ISBN: 979-8-9863055-4-7
Library of Congress Control Number: 2025924993

Marcia P. Fraser, Editor
Interior and Cover Design

On the front cover:
Eastview
Photo by Marcia Fraser

On the back cover:
Map of Williamson County by Michael Harris

Farm graphics courtesy of the Tennessee Century Farms Program,
Center for Historic Preservation, Middle Tennessee State University.

Williamson County Historical Society
P.O. Box 71
Franklin, Tennessee 37065
December 2025

To Judy Hayes

*For her active support and endless enthusiasm
in preserving Williamson County history.*

Williamson County Communities

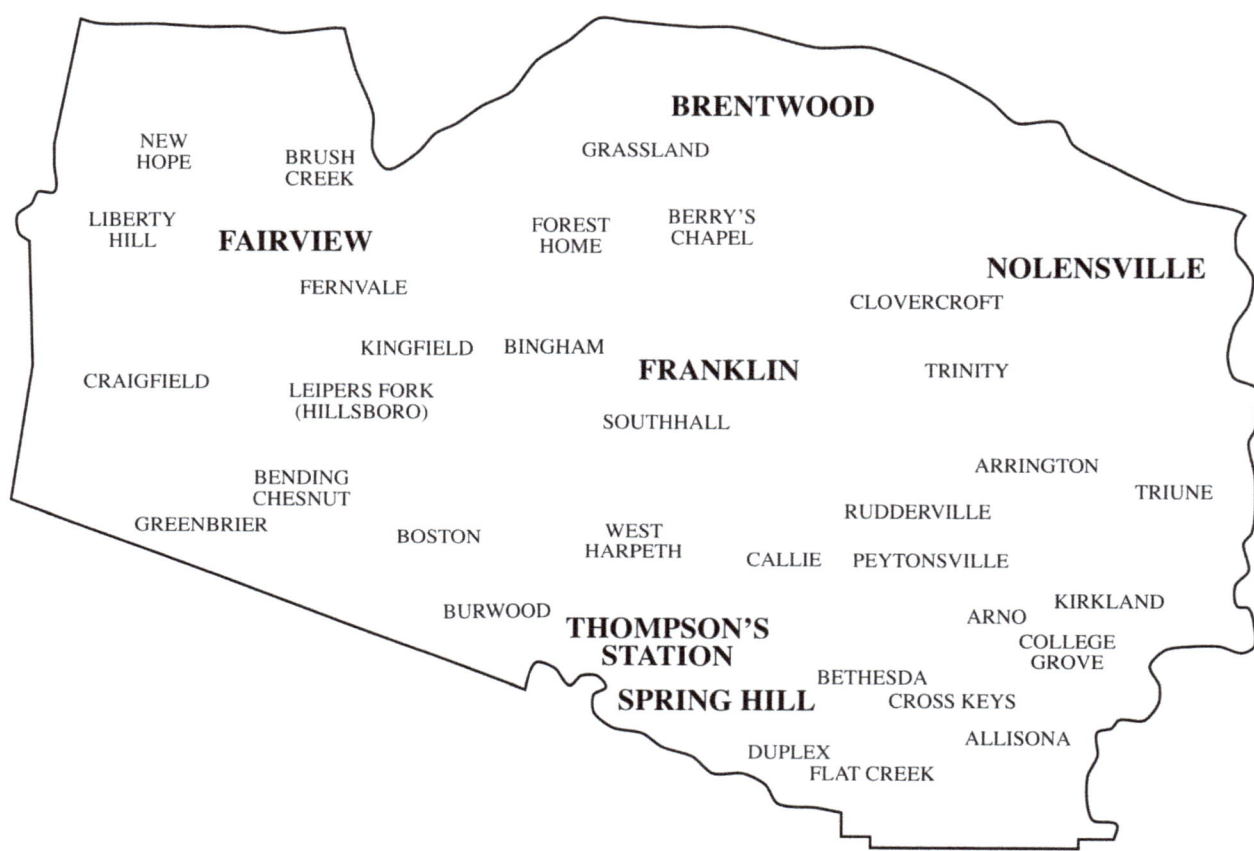

This map was drawn by Michael Harris of *That's Printing* and is based on the monument at Veterans Park in Downtown Franklin.

Contents

Acknowledgements	xi
Preface	1

SETTLERS COME TO THE BACK COUNTRY, 1785-1810

Glenn Acres, circa 1785	10
Maplewood, 1788	14
Beech Hill, 1796	20
German Farm, 1799	24
Locust Guard, 1800	29
Ozburn Hollow, 1806	33
Short Farm, 1807	35
Sherwood Green, 1808	39
Aurelia Acres, 1810	42

WILLIAMSON COUNTY BECOMES A LEADER IN AGRICULTURE 1815-1859

Poynor Farm, 1815	50
Woodland View, 1820	55
Blue Grass Farm, 1825	59
Valley View Farm, 1827	62
Cannon Farm, 1842	65
Hatcher Family Dairy, 1847	68
Bag End, 1848	73

Pleasant View Farm, 1848	76
Reams-Jefferson and Jefferson Farms, 1854	80
William Steele Farm, 1859	83

FARMING TRANSITIONS AS MODERN AGRICULTURE EMERGES 1869-1899

County Line Farm, 1869	91
Crystal Valley Farm, 1869	95
Bond Farm, 1870	97
Maple Crest Farm, 1870	99
Smith Brothers Farm, 1878	102
Sullivan Farm, 1881	104
Hunt-Beasley Farm, 1886	106
Johnsongrass or Lampley Farm, 1886	109
Charles Gentry Farm, 1887	112
Loreen and Mary Bond Farm, 1887	115
Cedar Creek Farm, 1888	117
Barker's Hillview Farm, 1891	120
Wilson Family Farm, 1893	123
Eastview, 1897	126
Long View, 1897	129
Pewitt Farm, 1897	132

OPTIMISM YOKED WITH REALITY, 1900-1922

Bud's Longview Farm, 1900	139
Walker Farm, 1900	142

Willow Run Farm, 1901	144
Hatcher Family Farm, 1903	146
Sullivan-Givens Farm, 1904	149
Peaceful Valley, 1905	152
Luster Farm, 1906	154
Bledsoe-Sullivan Farm, 1906	158
Nichols Farm, 1909	160
Penn Hollow Farm, 1913	165
Willow Creek, 1915	168
Thomas Jefferson Byrd Farm, 1920	172
Crestview, 1922	174

FORMER CENTURY FARMS REMEMBERED

Moss Side Farm, 1810	178
Rivers Meet Farm, 1816	179
Montrose Farm, early 1800s Midway Farm, 1832	180
Hillsdale Farm, 1842	181
Rodgerswood, 1853	182
Woodland Farm, 1857	183
Mockingbird Hill Farm, 1852 Reynolds' Grant Farm, 1865	184
Creekside, 1860s	185
Dripping Springs Stock Farm, 1869	186
Carter's Acres Farm, 1870s	187
Westbrook, 1887 Cartwright Farm, 1898	188
Cedar Lane Farm I, 1896 Cedar Lane Farm II, 1897	189

J.R. Givens, LLC, 1902	190
Selected Sources	191
Index	194
About the Publication Team	211

Acknowledgements

Without the cooperation of the Century Farm families, neither this publication nor the Tennessee Century Farms Program would be possible. This book is based primarily on materials shared by each Century Farm family in their original applications and subsequent updates, including the history of their farm, associated documents, photographs, and stories. To view the state's agrarian heritage from their perspective is a privilege and an ongoing learning experience. Their significant contributions to the county, to Tennessee, and the nation are appreciated and recognized.

This book originated as a proposal from Judith Grigsby Hayes, owner of Eastview Century Farm, and a tireless advocate for farmers and historic preservation. Not only does Judy suggest excellent ideas, she also made this publication a reality by generously sponsoring its printing. We are entirely grateful to you, Judy, for this latest gift you have provided to the people of Williamson County.

Rick Warwick, County Historian, supplied additional photographs from his collection housed with the Williamson County Historical Society, which is the publisher of this book. His numerous books and articles on county history were invaluable in its preparation. To converse with Rick, attend his presentations, or travel the county roads with him is always a seminar in local history. I join so many others here and elsewhere in thanking you, Rick, for your continuing efforts to collect, document, write, and publish to make sure we know and do not forget people, places, events, and material culture.

Marcia Fraser, recently retired as Special Collections Librarian for the Williamson County Public Library, served as the general editor and the designer of the publication. This is not an easy or small task and requires considerable time and specialized skill. Her knowledge of the county's history and people is exceeded only by her patience and her willingness to produce this book, the most recent in a long line she has designed and edited. These have added immeasurably to the body of knowledge and understanding of the county's heritage.

At the Center for Historic Preservation (CHP) at Middle Tennessee State University (MTSU) I am grateful for the assistance of my longtime former colleagues Dr. Carroll

Van West and Dr. Antoinette van Zelm. It was a pleasure to work with the Center staff and students over the years, and I am grateful for their continued support and interest. I join others across Tennessee and other states who appreciate the staff and graduate students of the CHP, who continue to carry on the important work of documenting, studying, and helping to preserve Tennessee's historic resources. Recognition also goes to the Albert Gore Research Center at MTSU, which now houses the Tennessee Century Farms collection.

Lynne Williams is the liaison between the CHP and the Tennessee Department of Agriculture (TDAG), where, among other duties, she is the Fair Administrator. To my knowledge, she has been involved with the Tennessee Century Farms Program longer than anyone. She was working at TDAG when the program began in 1975, and she continues to collaborate with others to ensure that new families receive their signs. Additionally, she attends Century Farm events at county fairs across the state. Thank you, Lynne, for your unwavering support of farm families and fairs, and for your gracious and timely support of many cooperative projects over the years.

The Williamson County Fair (WCF) has honored the Century Farms families annually with an exhibit and dinner. This, too, was an idea supported by Judy Hayes, who has served on the WCF Board for years. I appreciate being invited by the Fair Board and being supported and assisted by them in coordinating this effort for 15 years. During those years, it was my privilege to meet and visit with the county's Century Farm owners and their families, to recognize newly certified farms each year, and to present signs to those owners whose farms had progressed from 100 to 150 or 200-year signs. I shall always be grateful for that privilege.

My sincere thanks to family and friends who are supportive and interested in whatever project or publication I may be involved in. I am especially grateful to Larissa Barbee for her experience, skill, and willingness to assist with enhancing and resizing photographs. As well, recognition goes to Betty Merrill Webb, who provided assistance on several occasions.

To the farm families and to the readers who have chosen to learn more about agriculture in Williamson County through this book, we ask your understanding for images that did not reproduce as well as anticipated. Photographs came from different sources and vary in age, size, condition, and resolution. While efforts were made to adjust and produce the best images possible, some are still of a lesser quality. As well, sometimes the image was important and the only one extant, so it was used. We appreciate your kindness and charity, as we have done our best to give glimpses of each farm.

To my Skelley, Beasley, and Potts ancestors—all farmers in Williamson County for at least 200 years—I would not be here without you. Thank you!

All proceeds from this book will benefit the Williamson County Historical Society, which is celebrating its 60th anniversary in 2026, to support its publications and educational events.

From the volunteer team of Rick Warwick, Marcia Fraser, and myself, we are honored to bring this publication to you. We hope you will read and reflect on the history, the people, and the enduring legacy of these historic farms. They are among the most important places in the county. Then, with intention and gratitude, join with those who work to preserve farmland and support and respect farmers so they may continue to farm in Williamson County.

Caneta Skelley Hankins, 2025

County Line Farm, by Kristy Bergstrom

Preface

The Glass Mounds, located off Highway 96 West, are now part of the Westhaven community and golf club property in Franklin.

Within the span of history, European settlement in Williamson County is very recent. Little changed for millennia in this region of great forests, rivers, and natural springs that hosted wildlife and plants in unimaginable variety and abundance. Archaeological evidence and a few extant mounds and sites throughout the county tell of civilizations that flourished and disappeared. For example, the Glass Mounds on Highway 96 West in Franklin are rare Middle Woodland sacred sites, dating back around 2,000 years. The nomination of the Glass Mounds to the National Register of Historic Places by Dr. Kevin Smith of Middle Tennessee State University summarizes with "Glass Mounds are a resource of great significance for expanding our understanding of the prehistory of Williamson County and Middle Tennessee."

The latter Mississippian period is represented by the mounds and archaeological investigations that revealed stone box graves and artifacts ranging from 900-1450 A.D. at sites including Old Town on the Harpeth River near the Natchez Trace and the Fewkes Site. The Fewkes Site mounds on Moore's Lane, also the location of the 1830s Boiling Springs Academy, are a significant late Mississippian mound complex and village dating to approximately 1050-1475 A.D. The findings at each site reveal complex societies that teach us about their lives, the land, and natural resources. The continued preservation of these and other indigenous sites for future discovery and interpretation is a concern and a necessity.

Beginning in the mid-to-late eighteenth century, documentation by European adventurers, trappers, traders, and surveyors describes the land and the tribes, primarily the Shawnee, Creek, Choctaw, Chickasaw, and Cherokee, that were encountered in this region. Because of the wildlife, it was a fertile area for hunting and gathering, and vital to the survival of age-old traditions of tribes. As such, it was not always a peaceful existence because of factions that vied for the right to hunt and settle, even temporarily. Excavations at some villages reveal that these tribes annually planted corn, squash, and beans, often referred to as the "Three Sisters" because of their relative ease of growing and their sustaining food value. Squash grew on the ground, and beans entwined themselves in the cornstalks. These tribes shared a profound understanding, appreciation, and reverence for the land, its creatures, nature's ways, and cycles. Previous generations relied on the plentiful resources but left the land intact for those coming after them. This understanding of living in harmony with the land progressed for thousands of years.

Not until the late eighteenth century did Europeans and African slaves begin to enter this area and settle. In less than 250 years, this place, Williamson County, changed irrevocably. Most settlers farmed, and agriculture formed the foundation of the economy and society.

> The purpose of this publication is to bring to the forefront the significance of agriculture to Williamson County through the stories of its Century Farms–working farms owned by the same family for at least one hundred years. These farms represent every farm and farm family in the history of Tennessee and its number one industry—agriculture. Yes, agriculture remains the top industry in the state, contributing to the economy with a market value of agricultural products sold totaling over $5 billion annually. In Williamson County, the United States Department of Agriculture statistics from 2022 reported $339,881,000 in farm products sold at market from 1,153 farms. The number of farms and their acreage, however, dwindles each year, so this valuable contribution to the wealth of the county, at so many levels, is in jeopardy.

Edward S. Porter, Tennessee Commissioner of Agriculture, announced the Tennessee Century Farms program in 1975. (Image from the Tennessee Market Bulletin, October 1976, courtesy of TSLA)

The Century Farms Program originated with the Tennessee Department of Agriculture (TDAG) in 1975 as a way to commemorate farms and farming during the nation's bicentennial in 1976. Within two years, 630 farms had been certified. As more applications were received, TDAG invited the Center for Historic Preservation (CHP), established in 1981 at Middle Tennessee State University, to prepare a publication which was authored by Dr. Carroll Van West. The book included 783 farms and was published in December 1986.

From the early 1980s to the present, the CHP has administered the program, which now has over 2000 farms representing the state's 95 counties. The CHP and TDAG sponsored publications, a traveling exhibit, workshops, conferences, and sign presentations, as well as collaborative efforts with other agencies and organizations, to raise awareness of the importance of these remarkable treasures. Students and CHP staff continue to prepare nominations for farms to be included in the National Register of Historic Places, and other studies, including theses and dissertations, focus on Century Farms. The collection of applications, documents, and photographs is a remarkable compilation of agricultural history, paralleling Tennessee history from European settlement to the present. TDAG continues to supply the coveted yellow signs to new Century Farm families, depending on whether the farm is 100, 150, or 200 years old. They also work with county fair administrators, often supplying grants, to fund events that recognize existing and new Century Farms each year.

> To become a certified Century Farm, farm families are asked to prepare and submit a notarized application and documents providing proof that:
>
> 1. The farm was founded at least 100 years ago and has remained in the family.
> 2. The farm has at least ten acres of the original property.
> 3. The farm generates $1000 in annual farm income.
> 4. The farm has at least one owner who is a resident of Tennessee.

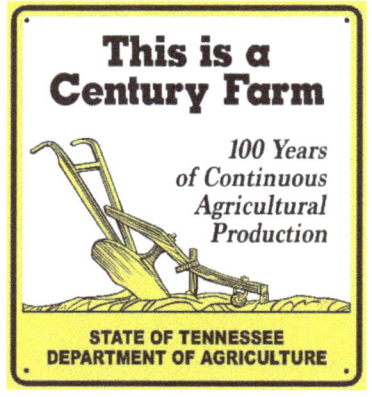
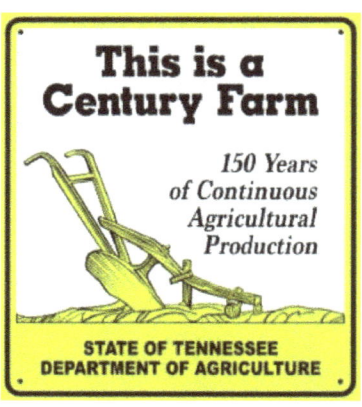
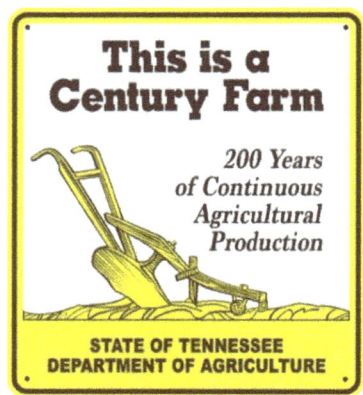

100 Years   150 Years   200 Years

It has been my privilege to be associated with Century Farms since the 1980s. I worked at the CHP from its inception in 1981 to 2013, when I retired as Assistant Director. For over a decade, one of my responsibilities was directing the Tennessee Century Farms Program. I traveled across the state, visiting many farms, each with its own valuable stories and contributions. What a learning experience that was, even for someone who grew up on a family farm. Let it be noted that every Century Farm is worthy of a book.

The entries in this book, however, are summaries. Likewise, selected photographs are intended to offer glimpses through the years of the people, the landscape, the buildings, and the animals. Accuracy is a goal, but one that is rarely achieved in a publication that deals with generations of names and dates from various sources. I offer my sincere apologies in advance to those families and readers who may find typos, errors, or misinterpretations of information.

To commemorate the 50th anniversary of the Tennessee Century Farms Program (1975-2025), this publication recognizes the Century Farms of Williamson County and the generations of each family and their singular contributions. Further, this publication brings the history of each farm to the present, when that information is available, and includes opinions and observations by working farmers on the challenges of farming in the county today.

Many residents without farming backgrounds do not realize that farming is a complicated profession, encompassing a range of disciplines including science, biology, math, computer technology, economics, mechanics, animal husbandry, and environmental awareness. With those skills is coupled a wisdom and stewardship of the land, a reverence for nature, and the desire to contribute to their community where they, like their forebears, have deep roots. Farmers deserve respect just as any other professional who has trained for years and practices their career daily.

Farmers are also the keepers of agrarian architecture, farming tools, equipment, traditions, and the rural lifestyle. The expanses of fields, pastoral views, and vistas that are such an integral part of the aesthetic scenery enjoyed and acclaimed by residents and visitors to Williamson County are primarily due to the work of farmers.

Alice Waters, a noted chef and pioneer of the farm-to-table movement, advises, "We eat every day, and if we do it in a way that doesn't recognize value, we contribute to the destruction of our culture and of agriculture. But if it's done with focus and care, it can be a wonderful thing. It changes the quality of your life."

To the generations of all farm families–past, present, and future–this book is offered with gratitude and with hope for a future that includes farms in Williamson County.

A family of goats on the Smith Brothers farm

**The American Farmland Trust reminds us:
"No Farms, No Food!"**

SETTLERS COME TO THE BACK COUNTRY, 1785-1810

Log cabin (Getty Images, whitemay)

I know of no pursuit in which more real and important services can be rendered to any country than by improving its agriculture, its breed of useful animals, and other branches of husbandman's cares. George Washington, Letter, July 20, 1794

The late 1700s charted a new direction of settlement when lone travelers and families began making the journey to the "back country" west of the Appalachian Mountains and the colonies. The North Carolina Land Act of 1783 offered 100 acres for $5, and many men and women found that bargain hard to resist. Though it was illegal to settle lands set aside by treaties to remain with Native American tribes, those tracts were increasingly encroached on and claimed by settlers each year. Confusion over boundaries, land grants, squatters, greed, and an unwillingness to abide by treaties resulted in hostilities between those idndigenous peoples and newcomers whose ways were a mystery.

For the Native Americans, their land, way of life, and culture was at stake and they learned they could not trust governments nor most of the people who moved into the territory. Members of tribes tried to abide by treaties, though the ownership of land was an incomprehensible concept. For those who had lived off the land since ancient times, the land was sacred, as were animals and plants. Survival was the primary goal of the tribes, and it became the primary goal of settlers who came in increasing numbers to acquire land, establish communities, build homes and barns, and farm. During the period between 1775 and 1800, three hundred thousand pioneers came into the Cumberland Valley. Many stories, documented and of oral tradition, describe the hostilities that occurred on both sides during this period of early European settlement as cultures and needs clashed. By 1793, with ever-increasing numbers of people moving into the area, the availability and use of firearms by settlers, and diseases that Native Americans had no immunity to, the contest was mostly over, tribes were pushed west, and small villages surrounded by farms took over the landscape.

In December of 1789, the state of North Carolina formally ceded its six westernmost counties, which had been formed between 1779 and 1784, to the Continental Congress. From this area, the state of Tennessee was formed and admitted as the 16th state in 1796. By this time, three farms featured in this section were already established. Designated as Pioneer Century Farms, they are Glenn Acres, Maplewood, and Beech Hill.

In 1800, the census recorded 2868 people in Williamson County, but by 1810, over 13,000 were counted. Immigrants, primarily from Scotland, Ireland, Wales, and England, first and second-generation colonials, Revolutionary War veterans, either taking up land grants or selling them to others, enslaved people, and a few free Blacks were part of the population. Williamson County was established in 1799. The transformation from wilderness to farms, crossroads communities, and towns had begun.

Settlers were risk-takers, and enslaved people had little choice but to come with their owners to this place with few outposts and well beyond what might be considered civilization. The land, however, was mostly fertile, water sources were readily available, wild game and fish were abundant, and wood and stone for building were plentiful. Most settlers believed that if they acquired land and worked hard, they could make a good life for themselves and for their families. For some, that was a dream that faded over time, but for others, it became a reality. For slaves, it would take much longer.

After one or two growing seasons in Williamson County, farmers knew what crops and livestock would thrive in the four distinct seasons. Some brought precious seeds, bulbs, and cuttings of plants, trees, and herbs from their former homes. Fortunate families were able to bring at least one milk cow, a mule or horse, and poultry. If not, they would purchase them as soon as they could upon arriving here. It was not long before enterprising merchants knew what necessities settlers would need, and businesses began to appear in small community settlements and places that would become towns.

Cutting trees and clearing land of vegetation and stumps was arduous and took time. The first plows were wood and not very effective unless the soil was quite rich, as it is in parts of the county. Turning plows with iron points came into use in the 1820s and dug much deeper furrows. Hilling was a planting option that yielded an abundance of food before land was readied for row crops. Most gardens included corn, squash, beans, turnips, and potatoes. Wild berries were picked until orchards could be planted and bear apples, peaches, plums, and pears. Wild mushrooms, greens, and herbs in the forest were available for the taking by those who knew what to gather. Any surplus food was dried or preserved, and fruits were not only dried but sometimes made into brandy. Distilleries were common on individual farms. Wild honey was a treat when sugar was not available, and settlers made beehives and captured bees to have a good supply. Fish and small game, as well as deer, bear, and elk, fed families year-round. Though the landscape was already changing, it sustained those who took time to understand the ways of nature and work with it season after season.

While early farmers were learning what would grow here and how best and when to grow it, they were also occupied with building shelters for themselves and their domestic animals. Most of the first dwellings and outbuildings were constructed from logs. Trees were everywhere and a building, crude though it might be, could be built with relative ease and just a few tools. Families coped daily with growing vegetables, hunting, and fishing to have enough to eat while clearing fields, refurbishing, and making clothing. Their primary goal was to survive, and they did not likely dwell on the fact that they were establishing communities, familial lines, agricultural traditions, and constructing buildings that would remain intact, some into the twenty-first century. That, however, is what European settlers, enslaved people, and the Century Farm families featured in this section did when they came into the "back country" that is now Williamson County.

SETTLERS COME TO THE BACK COUNTRY, 1785-1810

GLENN ACRES, CIRCA 1785

MAPLEWOOD, 1788

BEECH HILL, 1796

GERMAN FARM, 1799

LOCUST GUARD, 1800

OZBURN HOLLOW, 1806

SHORT FARM, 1807

SHERWOOD GREEN, 1808

AURELIA ACRES, 1810

Glenn Acres, circa 1785

Remodeled and added to by generations, the log house within Glenn Acres' farmhouse was constructed by Isaac and Mary Ann McGuire Gillespie in the late 1700s.

The Flat Creek community, located in the southeastern section of Williamson County, takes its name from the meandering stream that attracted animals and later Native American tribes who passed through the area to hunt and fish. When families made their way into the wilderness of what would be southeastern Williamson County in 1799, a community formed around this reliable source of water for people and livestock.

Glenn Acres, often referred to as Gillespie Place, is one of the few Century Farms that were established before statehood was approved in 1796. A designated Pioneer Century Farm, it was owned by Thomas Gillespie before the United States Constitution was written and ratified in 1787-89. Gillespie and his wife, Naomi, lived in Rowan County, North Carolina, where they owned a large plantation. Gillespie was a neighbor of and served as a commissary to General Griffith Rutherford, for whom Rutherford County, founded in 1803, is named. A grant of about 4,000 acres was issued to Gillespie following his application in 1783, and his land was surveyed in 1785. Thomas and Naomi Gillespie did not, however, come to claim his grant, for both died in December of 1797.

James K. Polk, 11th President of the United States, was a descendant of Thomas and Naomi Gillespie of North Carolina. The land grant issued to Thomas is the basis for one of the oldest Century Farms in Tennessee – Glenn Acres. Polk would visit his relatives when traveling from Columbia to Murfreesboro for work and to visit Sarah Childress who would become his wife.

Their son, Isaac, who also fought in the War for Independence, and his wife, Mary Ann McGuire Gillespie, arrived at what is now Flat Creek shortly after their parents' death and settled on approximately 600 acres inherited from his father. They built a log house, which is intact within the current farmhouse. From Isaac's will, we learn that he and Mary Ann had six daughters and four sons. The family raised sheep, cattle, swine, horses, and grains. We also know from his estate settlement that the family had slaves, and three are mentioned by name – Sally, Jim, and Bur.

Isaac's brothers, David, Thomas, and George, also came to Tennessee to assume ownership of land inherited from their father. Their sister, Lydia, was married to James Knox. It was their daughter Jane who married Samuel Polk of Maury County and was the mother of James Knox Polk, the 11th President of the United States. Family tradition recalls that James K. Polk would visit his uncle and family on Flat Creek on his frequent journeys between his home in Maury County and Murfreesboro for work and to visit his future wife, Sarah Childress. At least three of the Gillespie brothers, along with other Revolutionary War soldiers, are buried in the Moses Steele Cemetery.

William "Bill" Henry Gillespie acquired a portion of the original land grant in 1866. The son of David and Anna Gillespie, Bill married Elizabeth Reed. Their sons were Samuel and Wallace. Samuel inherited some of the acreage in 1914. For over 30 years, Samuel and his wife, Pauline Stephens Gillespie, lived and worked on the farm with their children.

In 1948, the farm was acquired by Martin Dodson "Jackie" Glenn and Lola Reed Glenn, relatives of the Gillespies. Mrs. Glenn, who married Bill Bowersock after Jackie's death, was a teacher in Bethesda and, in 2002, was inducted into the Tennessee Teachers Hall of Fame. She was very active in the Home

Demonstration Club, Flat Creek Community Center, and Bethesda United Methodist Church, among other community groups and organizations.

Calvin Glenn, son of Jackie and Lola, became the owner of 121 acres of the original farm in 1979. He and his wife, Sandra Thompson Glenn, have raised tobacco, hay, corn, soybeans, wheat, and beef cattle through the years. They are the parents of Tony, Jackie, and Susan. The Glenns live in the house that includes the log building constructed by the pioneering Gillespie family. Glenn Acres is the oldest certified Century Farm in Williamson County and one of the oldest in the state of Tennessee.

The multi-use hay and stock barn at Glenn Acres is typical of barns built across the county in the 1940s and 1950s.

Lola Glenn Bowersock was a well-known and respected teacher, an active member of the community, and the matriarch of her family.

Jackie and Lola Glenn with Janice and Calvin

Lola Glenn Bowersock

The Old Hen House, as it is called by the family, is a reminder of the long tradition of raising poultry for eggs, meat, and to market on most farms. Many farm wives used the money from selling eggs and chickens to supplement the farm income.

Gathered in the old house are members of the Calvin and Sandra Glenn family. **Pictured from Left:** *Emily Kester, Lola Bailey, Misty Bailey, Madelyn Smith, and Brady Perez.* **Back Row from Left:** *Tucker Glenn, Mason Kester with Evan Bailey in his lap, Calvin Glenn with Blake Perez in his lap, Sandra Glenn with Stella Kester in her lap, Brandon Jaworski, and Shelby Herbert.*

Maplewood, 1788

Maplewood, built in 1837

The Lee Family property in the Duplex community has been featured in many publications and is listed in the National Register of Historic Places for its early settlement, buildings, and agricultural significance. It is the second of three Century Farms designated as a "Pioneer Century Farm" for it was founded before Tennessee became a state.

Genealogy and family history often raise questions, particularly for families that own property from this early period of settlement. Names that reappear in succeeding generations further complicate family history. The story of Maplewood is complex, but the current owner, John Lee, has sifted through family papers and legal documents to trace and resolve the farm's initial and subsequent lineage.

The farm's history begins with a Revolutionary War Land Grant (No. 606) awarded to James Polk by the State of North Carolina, for which Polk received the certificate on July 10, 1788. The 5000-acre tract was described as "on the south side of a branch of the Duck River." Charles Polk acquired the acreage sometime after 1788 from his relative, James, who may have been his brother. Mary, the sister of Charles Polk, married Daniel Brown of Charleston, South Carolina, who purchased the entire tract from his brother-in-law. This purchase was prior to 1802. In that year, Brown died, though his estate was not settled until 1808. Brown's will directed that the property, now in the new state of Tennessee, be left to his brother-in-law, Samuel Lee of Salisbury, Connecticut, who was married to his sister Elizabeth Brown Lee.

Samuel Brown Lee (1798-1865) traveled from Connecticut to Tennessee in 1816 to settle on land that the Lees have lived on and farmed since that time.

Samuel Lee registered the deed to this property in Williamson County in 1810. In his registration, Samuel Lee quotes from Daniel Brown's will which had this directive, "I give the tract of 5,000 acres of land lying in the State of Tennessee which I purchased of Charles Polk for which I have his obligation to make title to Samuel Lee of Salisbury on Conn restricted for the use of his children by my sister Elizabeth." The deed goes further with this explanation, "being part of the tract of 5000 acres originally granted to James Polk by Grant bearing the date the tenth day of July in the year of our Lord 1788 under the seal of the state of North Carolina." (page 300 of the Williamson County deed book, which references page 200 of Book B). Samuel and Elizabeth Brown Lee lived in Salisbury, Connecticut, and had nine children. When Samuel Lee came to Williamson County in 1810 to register the deed for his land, he sold 2000 acres. Samuel Brown Lee, Samuel and Elizabeth's fourth child, came to live on the 3000-acre farm in 1816, and this began the Lee family's residency for more than 200 years.

According to the family, who consulted log-building experts, a single-room cabin on the property was built between 1780 and 1800. Another single pen with a connecting dog trot was added between 1820 and 1830. Family tradition indicates Samuel Brown Lee worked for a dry goods store in Spring Hill and is thought to have purchased it a few years later. It is known that he lived in Memphis for a time because letters between him and his future wife, Susan Napier, are still in the family's possession. Susan was the daughter of John Napier, whose name and influence are recognized in the history of Tennessee's lucrative, far-reaching iron industry that emerged before the Civil War. Lee worked for Napier in some capacity and learned enough about the business to venture into it with his cousin, James Gould. They built and operated the Lee and Gould Furnace in Hickman County, which employed hundreds of workers. The furnace was in blast only until 1835 because of the scarcity of high-quality iron ore in that area. The impressive stone stack remains near Sugar Hill and is noted with a marker.

John Wills Napier Lee is seated while his granddaughter Eunice stands next to him. Standing behind are his children, Bess, John, and Eunice Lee Mallory. The backdrop is the ca. 1837 farmhouse, which is the main residence today.

By 1837, Samuel Lee was cultivating cotton and raising livestock on the farm where he constructed outbuildings and the two-story farmhouse. He married Susan Napier that year, and their descendants have occupied the house since that time. By 1850, the farm was valued at $15,000, and mules were a good portion of that total. Lee and a slave drove the mules to Mississippi for sale. At least twenty-six slaves were owned by Lee in 1860, and his farm's worth had doubled since the previous census. By this time, Samuel was a widower for Susan died in childbirth in 1850. He relied on Mammy Ann, a slave, to help raise his four surviving children.

Before 1860, the original 3000 acres were divided and sold to benefit Samuel Lee and the Lee siblings, who never came to Tennessee. At that time, Lee owned 600 acres and began transitioning from cotton to tobacco as his primary cash crop. He also increased production of wheat and corn, marketing thousands of bushels of each. The Lee farm's location and its productivity made it vulnerable to foraging by soldiers on both sides during the Civil War. Not only were stores of grain and most of the livestock taken, but fences were torn apart and burned for firewood while the house was used as a hospital. Two sons, John Wills Napier Lee and Samuel Brown Lee, Jr., enlisted in the Confederate Army on September 28, 1861, at Carnton. Their unit adopted the name "The Williamson County Cavalry." John Wills Napier stayed with this unit throughout the war. Samuel Brown Lee transferred to the artillery midway through the war. With the end of the war in 1865, the sons returned home to a farm that was mostly destroyed, arriving not long before their father died.

The farm was then divided among the surviving children, Samuel Brown Lee, Jr, John Wills Napier Lee, Charles Alfred Lee, and Florence Amanda Lee. Three of Samuel and

Susan's children attended college. Samuel Brown Lee, Jr., graduated with a degree in engineering from Cumberland University. John Wills Napier Lee suspended his studies at Cumberland University to enlist in the Confederate Army. Florence Amanda attended the new and highly respected finishing school for ladies, Ward Seminary in Nashville. The school was established by her mother's cousin, Dr. William E. Ward, and his wife, Eliza Hudson Ward. This institution later became Ward-Belmont College and is now Belmont University.

With the help of former slaves who continued to live and work as tenant farmers, the Lees began to rebuild the farm. They raised corn, wheat, and cotton, but not tobacco. Poultry was important, too, and John is noted for collecting about two hundred dozen eggs in 1870. The brothers began to revive the family tradition of raising mules and horses. Charles had five horses and seven mules, and John claimed thirteen mules and six horses. A horse John acquired in 1872 brought national and international fame to Maplewood. "Duplex" won the World's Championship Pacing Record in Detroit in 1887 and became a sire of thoroughbreds across the nation. Samuel Brown Lee was elected to the state senate in 1897.

John and Mona Lee on the front steps of their home

Succeeding generations of Lees continued to farm and be an active presence in the community through the twentieth century. They went from dairy cattle to beef cattle and raised tobacco in the first half of the century. Charles Alfred Lee acquired Florence Amanda Lee Farrell's portion at her death, and John Wills Napier Lee acquired Samuel B. Lee's tract of the farm at his death. After John Wills Napier died in 1921, his only son, John Wills Napier Lee, Jr., acquired his father's parcel from the estate, and in 1944, he acquired the acreage of Charles Alfred Lee from his estate. With the death of John Wills Napier Lee, Jr. in 1964, his two sons, John Wills Napier Lee, III, and Sam L. Lee, inherited his portion of the farm.

John and Mona Lee's children and their families are the seventh and eighth generations of the Lee family to call Maplewood home. Left to Right: Martha Ernst, Eli Ernst, Christian Ernst, Laura Lee Ernst, John and Mona Lee, Mason Lee, Jack Lee, Megan Lee, and Miller Lee.

In 1985, John Napier Lee purchased his uncle Sam L. Lee's share of the farm. In 1990, the sixth generation, John Napier Lee and Nancy Anderson Lee inherited the farm from their father, John Napier Lee, III, whose wife was Martha Broyles Lee. John Napier Lee and his wife, Mona Robertson Lee, raised their children on the farm and continue to be the caretakers of the 1837 house, the outbuildings, the Lee family history, the cemetery, and the land.

In 2025, Maplewood has approximately 600 acres. In its long history, it was divided into fourths, then into halves, and is now once again at its 1860-acre size, owned by the family. About 80 acres are in row crops and a commercial cattle operation. Following family tradition, Maplewood is home to three generations today. In 2024, two new houses were built on Maplewood. Family currently living on the farm and owning portions are John and Mona; Nancy Lee Anderson; Laura Lee Ernst and her family, including Christian, Eli, and Martha Ernst; and John (Jack) McCutcheon Lee, Megan Miller Lee, and their children, Miller and Mason Lee. Maplewood's history is long and well-documented, and the current owners and residents add to that history each day.

Duplex helped establish Maplewood Farm's reputation and named a community. The famous horse won the World's Championship Pacing Record and sired many thoroughbreds. This handbill from 1896 may be the only extant image and description of Duplex.

(Photo credits for the John and Mona Lee family and house photographs: Meredith at Vintage Tree Photography)

Beech Hill, 1796

Within this comfortable farmhouse is the original log home of the founders of the farm, who came to what is now the College Grove area in the late 1700s.

Beech Hill is one of two Century Farms in Williamson County designated as Pioneer Century Farms. Historically referred to as the Ogilvie Farm, this property in College Grove has been in the same family for as long as Tennessee has been the 16th state.

William and Mary Harris Ogilvie journeyed across the Appalachian Mountains and then traveled south from Nashville to acquire land and build a log house near a spring that continues to supply water for the farm to this day. As William expanded and improved his property, he gave and sold parts of his landholdings to his sons. In his 1813 will, he gave to his son Richard 315 acres, including the house, cabins, and farm buildings. When Richard died in 1822, he willed the farm to his wife, Cynthia, and their youngest son, James Smith Ogilvie. During this period of early settlement and until the Civil War, slaves also lived and worked on the farm. One slave cabin (circa 1830) remains intact as a reminder of the men, women, and children who lived in these dwellings.

The stone springhouse, one of the original buildings dating from the late 1700s, covers the source of water that has never run dry. It was also the place to store milk and other perishable food until refrigerators were invented and available to farm families.

James Smith Ogilvie and his wife, Rachel Webb, raised six children on the farm. In 1897, when James died, he left property to their children, and his sons, Samuel Jason Ogilvie and James Smith Ogilvie II, purchased their sisters' shares. The two brothers farmed together for a time but eventually divided the acreage. Samuel retained the homeplace with most of the buildings and 150 acres, while James built a new house for his family on 165 acres.

Samuel Jason Ogilvie died at the age of 36, leaving three young children for his widow Anna Rucker Ogilvie to raise. With the help of African American families living on the property as tenants, she was able to keep the farm for her children, James D., Rachel, and Samuel. During World War I, James, a Marine, served in Europe. According to the family, he was given the job of caring for a team of mules and an ammunition wagon as the troops made their way through France and Germany. After he returned from the war, James became responsible for the farm's operations. Rachel and Samuel decided to leave the farm, and Bettye Maxwell Ogilvie, their sister-in-law and wife of James, purchased their interest. This couple operated the farm for the next three decades, raising their children, Samuel Rucker and Elizabeth Maxwell, in the original house, which Bettye had updated. She also took special pride in maintaining the family burying ground that was established in 1807. James and Bettye named the farm "Beech Hill" because of the number of beech trees that grew on the place. The house, spring house, and slave cabin are listed in the National Register of Historic Places.

Tobacco was a main crop for decades, and the existing barn is a fine example of that specific construction. James grew tobacco from seeds he planted each year in beds, and day helpers were hired to plant, cut, and hang the tobacco. After it was stripped in the late fall, he hauled it to the tobacco warehouses in Franklin, where it was sold. While the barn ceased to be used for that crop after the tobacco buy-out of 2004, it was used to store hay.

Samuel R. Ogilvie and Elizabeth Ogilvie Battle, the 3rd-great-grandchildren of the founders, received the farm's 150 acres in 1964. In 1992, Elizabeth and her husband, William Robert Battle, became the sole owners. Elizabeth continues to make her home on the farm in the house that includes the log house built by her ancestors, William and Mary Harris Ogilvie. Rob Battle and his family live on the farm, while Valerie Battle Kienzle appreciates the heritage of the farm and photographs the buildings and landscape on her frequent visits.

Currently, the land is rented to a local farmer who raises Black Angus cattle on the lush pastures. Beech Hill Farm, located one mile south of College Grove, has been featured in many publications and is a landmark property not only in the community and county, but in Tennessee, with which it shares a birth year – 1796.

Tobacco was a cash crop of the Ogilvies for many years. Barns like theirs were built for tobacco production. In today's dwindling tobacco market, many barns have disappeared.

BEECH HILL, 1796

This rare slave cabin at Beech Hill is a reminder that many farms depended on the work of enslaved men, women, and children.

From Horton Highway, the entrance to Beech Hill is marked by its 200-year Century farm sign. The farm remains in the Ogilvie family after nearly 230 years.

Elizabeth Ogilvie Battle, Rob Battle, and Valerie Battle Kienzle are the current owners of Beech Hill, also known as the Ogilvie Farm.

German Farm, 1799

The 1880s frame farmhouse, photographed in the 1950s, evolved over the years to accommodate the family's changing needs.

Northeast of Franklin and enveloped by neighborhoods is a working farm that traces its heritage back to the end of the eighteenth century. One of only three farms in the county that dates before 1800, the German Farm's current owners acknowledge Daniel German and Fanny Puckett German as the ancestors who first farmed this land in 1799. Their son Zacheus, at the age of eighteen, married Emaline McEwen, who was sixteen. With their union, the young couple joined two of the earliest families who settled in the county, and numerous descendants of their fourteen children remain residents. On 300 acres, the Germans bred horses and raised grains, fruits, and livestock for the family and for sale. Slaves worked on the farm and in the substantial two-story brick house. Virginia Bowman writes in her entry for the German house in *Historic Williamson County* that as Federal troops approached the farm during the Civil War, the invalid daughter, Sarah German, was cared for and hidden by slaves who were charged with her safety.

GERMAN FARM, 1799

The house was a backdrop to this family photograph of Henry Mortimer Williams and Cynthia German Williams and their children. The young boy in the photograph is Oscar F. Williams, Sr. Oscar was born in the house, as was his son, John Williams, Sr. Today, the great-granddaughter of Oscar and granddaughter of John, Sr., MaryLynn Williams, makes her home in the historic dwelling.

When the German Farm was certified as a Century Farm, at least 40 acres of the original farm were still in use. They were owned and operated by brothers Oscar Fitzallen Williams, Jr., and John R. Williams, Sr., the great-grandsons of Zacheus and Emaline German. The Williams brothers, on a total of 480 acres, managed beef cattle, and annually grew row crops including tobacco, hay, corn, wheat, and later soybeans. They also provided custom harvesting throughout the county and for surrounding area farms. John served in the U.S. Army Corps during World War II and was the eighth generation of his family to live and work on the farm. He was married first to Mary Armenia Gillespie and then to Talitha Whitley. After his brother Oscar's death in 1986, John continued the operations of the farm, along with his son, until his death. The family recalls John Sr. enjoyed whittling and made many pocket-sized crosses, which he delighted in giving away. Mr. Williams was a member and deacon of Franklin's First Baptist Church.

John Williams, Sr. and John Williams, Jr. worked together and are the seventh and eighth generations of their family who have lived on and farmed the German Farm since 1799.

John Williams, Jr., grew up farming with his father and uncle and continues that tradition today. He and his wife, Janice, became the owners of 80 acres in 2018. Their children, MaryLynn and John Joseph "Joey," were raised on the multi-generational farm. Drawing on the experience and knowledge of his father, grandfather, and great-uncle, Joey pursued farming as a science and has a Ph.D. in Agronomy from Mississippi State University. He is the regional Crop Protection Research Manager for Bayer and is also engaged in research with the Southern Weed Science Society. He assists his father in making decisions on crops and cattle production. His wife, Sarah, has a master's degree in animal nutrition from Mississippi State University and helps with beef cattle and hay production. MaryLynn is completing her master's degree in education and teaches third grade. She lives in the house once occupied by her grandparents and previous family generations and helps with harvesting hay and monitoring cows during calf season.

Mr. Williams reports that the family is currently raising beef cattle, primarily Herefords with some Angus. For the first time in years, they raised alfalfa last year at the request of horse owners, but he explains, "only a small amount and to test the market and crop." The other fields are in fescue and orchard grass mix with monitored weed control for cattle and, more recently, for local horse farms because hay is "hard to come by in winter months due to the increasing loss of farms in the area."

When asked to address the biggest challenges facing farmers in Williamson County, Mr. Williams reinforced what other farm owners had said about the dangers of moving equipment and hauling hay due to traffic. As neighborhoods surround farms and residents are either unaware or heedless of their impact on adjacent farmland, a lack of knowledge and respect for the ways of agriculture is a fundamental problem. For example, Williams noted that passersby "throw garbage into fields that are seeded, fertilized and sprayed for weed control," all of which is extremely expensive. Fencing must be installed and maintained for the safety of both animals and people. Mr. Williams reported that some neighboring residents have moved or torn down fences on his property, allowing animals to wander out of the safe pasture onto roads. He also cites the increasing cost of labor and the scarcity of people familiar with agricultural practices as yet another issue that must be addressed. Furthermore, some federal, state, and local regulations, such as restrictions on killing predators of livestock and row crops, are complicated for farmers. The constant pressure to sell to developers and build more homes in the county is common to every farm family.

Despite these daily and seasonal challenges, Mr. Williams writes of his ancestral farm and his livelihood, "It's my heritage, it's my soul, it's my memories of hard but good times. It's what drives me to fight for my kids. No amount of money can give me the same feeling as when I see my kids with me doing the hard stuff, just like my ancestors before me. I can feel looking over my shoulder and saying, 'Good job, keep it up.' I pray that my kids feel it too."

The German Farm, with its daily and seasonal work, cattle, and landscape, is a rare survivor of the productive farms that once thrived in this area.

A new tractor is a large investment for farmers, no matter when it is purchased. John Williams, Sr., gives a ride on a pristine Allis-Chalmers to his grandson, Joey. More recently, the purchase of their first cab tractor is cause for a family photo. From left are MaryLynn, John, Jr., Janice, and an older Joey with his wife, Sarah.

Locust Guard, 1800

Bob and Charlene Ring are the caretakers of Locust Guard and active citizens while raising their family to appreciate the farm and the county. Bob served as county commissioner and was county executive for 16 years.

The Grassland community, north of Franklin, is a vibrant community with schools, churches, and many residences. Because of its proximity to Davidson County, primarily via busy and sometimes congested Hillsboro Road, an early turnpike, this area has always been a convenient and beautiful part of the county in which to live. The land is fertile, springs and water sources are abundant, and herds of livestock and fields of row crops were once commonplace along the roadways. Locust Guard is one of the few remaining working farms, and its history, even to the present day, is bound to that of Williamson County in many ways.

Locust Guard is a rare survivor in this part of the county due to the diligence and agricultural knowledge and practices of generations of Motherals and Rings. John Motheral, a Revolutionary War veteran, and his wife, Jane Currie, came from North Carolina and established a 400-acre farm in 1800 near the Harpeth River. By the time their son Joseph and his wife Anness became owners, they had made significant changes and increased the farm's productivity. The house was built, and dry stone walls were constructed; a grist mill, barns, milk house, apple house, and weaving house became part of the farmscape. Corn, wheat, and other grains, along with dairy and beef cattle, sheep, and swine, were raised for family use and to sell at market.

In 1858, Emma Tennessee Motheral, one of three daughters who inherited property from their father Joseph, married Henry Eleazar Ring, a professor whose own lineage could be traced to the Mayflower. The two were not married for long when Henry died, and Emma returned to Locust Guard to live and raise her two young sons.

The seven children of Henry Ring and Fannie McClellan Ring are, from left, Frank, who died in France during WWI, Beth, Emma Mai, James E., Andrew, Ned, and Jasper. This photo dates from circa 1913-1914.

It was the son named after his father who became the next owner and operator of the farm. Henry married Sarah Frances McClellan, and they had seven children. Henry was a graduate of the University of Tennessee and not only farmed but also served as a county commissioner. Honeybees, nuts, and alfalfa were added to the farm's products. All their sons attended the University of Tennessee. Jasper and Ned played football at U.T. Later, Jasper achieved the rank of Army colonel and worked as an engineer on dam projects in the western states. The family grieved for Frank, who died in France during World War I. The youngest son, Andrew, studied engineering at U.T. and began his successful career in radio. Ned and James E. returned to manage the farm and to construct new buildings and install an irrigation system. James expanded the peach orchards, planted pecan trees, and was President of the Tennessee Horticulture Society. The Ring family believed their daughters also deserved an education, and Beth and Emma Mai became schoolteachers. Emma Mai taught for 26 of her 39 years in the classroom at nearby Grassland School.

Robert Ring was born in Washington, D.C., where his father, Andrew, was called to help form the Federal Communications Commission (FCC). From his earliest days,

Robert visited and worked with his aunts. uncles, and cousins on the family farm north of Franklin during the summer months. "I could not get enough of it," he remembers. Following family tradition, Robert attended the University of Tennessee, graduating from the College of Agriculture. After receiving his degree, he moved to the farm and joined his Aunt Emma Mai and Uncle Ned in farming. Following Ned's death, the farm was deeded to Robert and his aunt in 1961. Emma Mai was a driving force in keeping the farm intact and told her siblings, "Boys, you will cooperate." Gradually, Andrew bought the interests of other heirs, and with the deaths of Andrew and Emma Mai, Robert and his wife, Charlene, became the owners of Locust Guard.

Emma Mai Ring, who lived at Locust Guard most of her life, was a teacher at Grassland School. She and her nephew, Robert Ring, became the fifth and sixth generation owners of the family farm.

Robert and Charlene Ring have lived on this farm on Moran Road, just off Hillsboro Road, since 1960. Robert remembers the close-knit farming community he knew as a youngster when visiting the farm. The Rings raised their family on the farm and still live in the circa 1823 house built by his ancestors. From this vantage point on Moran Road, they have witnessed remarkable changes all around their farm. Robert, former Williamson County Commissioner and the County Executive for 16 years, has a unique perspective.

When Ring became County Executive in 1982, he sold his livestock and arranged for a neighbor to put in crops on shares. That plan continues today, and Ring explains that "Eddie Saunders with his large modern machinery is able to crop our mostly floodplain crop land with a profitable outcome." Bob explains that "It has never been my intent to sell any part of the farm as I have felt more of a caretaker than an owner." Planning for the future, the Rings have considered that other farms had to be sold to pay inheritance taxes, so they have given their children undivided shares of the acreage. When they become full owners, Ring acknowledges that "they are free to make decisions as current law and their circumstances dictate."

Locust Guard is a landmark on Moran Road. The stories and legacy of the Motheral and Ring families are an affirmation and representation of a progressive agricultural heritage that has sustained Williamson County throughout its history and into the present.

Robert and Charlene Ring's grandchildren, Edie May and Daniel, are first cousins and represent the eighth generation of the family.

Left: Dating to no later than 1823, the log milk house, built above a stone cellar, is possibly the oldest building associated with dairying in the county. The milk house, a log smokehouse, and the original log house within the current dwelling are survivors from the county's early history. Above: Row crops have long been the primary commodity of Locust Guard, with its fertile bottom land along the Harpeth River.

Ozburn Hollow, 1806

Standing in front of the still-extant two-story log house that has been covered with weatherboarding in this early photograph are three generations of the Ozburn family: Ellen, Willie Dean, Alma, John, Sr., Mildred, Uncle Will, Jack, Dessie, Ruth, Leslie, and an unnamed farm hand. Ozburn Hollow is protected from development through a conservation easement with the Land Trust for Tennessee.

Near the Williamson and Rutherford County lines is an 1806 farm that is protected from development because its owners, F. Perry Ozburn, Jr., and Elaine Ozburn, placed 480 acres in a conservation easement with the Land Trust for Tennessee in 2006. The Williamson County Historical Society also noted the significance of this early property with a wayside marker that summarizes its nearly 220 years. Other publications have also documented Ozburn Hollow.

The grave marker of farm founder, Jane Wylie Ozburn, in the family cemetery.

The grave marker of farm founder, Robert Ozburn, in the family cemetery.

Robert Ozburn belonged to one of thousands of families from Scotland, England, Ireland, and Wales that came to North America in the latter part of the eighteenth century. Born in York County, Pennsylvania, in 1755 to Scottish immigrants, he was in North Carolina by 1775 when he enlisted to fight in the Revolutionary War. Robert married Jane Wylie of North Carolina in 1785, and with their children, they relocated to a large property in Williamson County in 1806. A log house, spring house, smokehouse, and log barn remain from this initial period of settlement. After Robert's death, Jane applied for and received a widow's pension because of her husband's service. A nearby cemetery includes burials of the founding couple and several of their descendants.

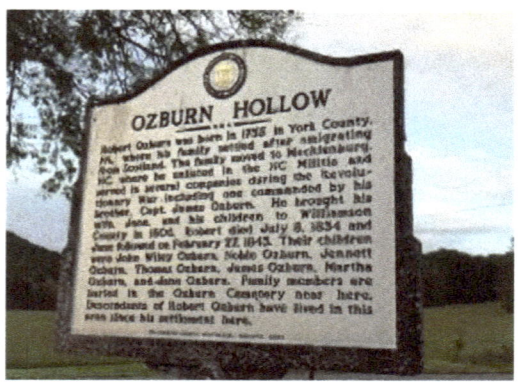

Generations have farmed this acreage, contributing to the history of the county and state. Frank Ozburn founded Ozburn-Hessey Logistics in 1951 and built it into a national and international enterprise. His son, F. Perry Ozburn, Jr., continued to lead the company for several years before it was sold in 2015. Mr. Ozburn explained in an article in the *Williamson Herald*, dated November 29, 2012, by Carole Robinson, *that he wanted to ensure* the land remains in his family for his three sons and granddaughters. He concluded with, "You want to make the right decision. You want it to stay in the family. The Land Trust option helped solve the problem. Now my grandkids will always have a place to stay." Ozburn Hollow will remain a landmark farm because of the foresight of the eighth generation.

Short Farm, 1807

The restored circa 1810 log house built by William Irby and Margaret Anderson Boyd is located across Boyd Mill Pike from the Short Farm.

West of Franklin, on ever-changing Boyd Mill Pike, the Short Farm traces its lineage to the Boyd family for whom the road, and the once well-known mill, are named. William Irby Boyd and Margaret Anderson Boyd of Halifax County, Virginia, came into Williamson County in 1807. Starting with 117 acres, Boyd purchased 76 acres in 1814 and added another 315 in 1818. In the Bingham community, located along the West Harpeth River, the farm is fertile and well-suited for both crops and livestock. The original two-story home of the Boyd family was wisely built on a rise above the river, which is known to flood.

William Anderson Boyd, the second owner of the farm, also acquired Boyd Mill and Boyd Distillery in Still Hollow.

William Anderson Boyd, son of the founding couple, expanded the 50 acres accorded him in his father's will to almost 1000 acres. Engaging in other enterprises, he was a stockholder in the South Harpeth & Franklin Turnpike, which is the early and official name of Boyd Mill Pike, and he owned Boyd's Mill and Boyd's Distillery in Still House Hollow. The 1860 slave schedule lists ten slaves who worked to produce timber, grains, livestock, as well as vegetables and fruits. Some of the men likely worked at the mill and at the distillery. A log house with a loft and half-dovetailed notching was likely a slave dwelling and later a tenant house. Boyd was involved in the building of the Nashville-Hillsboro Turnpike (Hwy. 431) after the Civil War.

Tennie Boyd inherited at least 50 acres from her father, William Anderson Boyd, in 1881. Married to Benjamin F. Short, the couple continued to raise grains, sheep, cattle, swine, horses, and mules. Their son Jesse E. Short, Sr., inherited the farm from his father in 1906. Jesse's children were Jesse E. Short, Jr., James Cotton Short, and Lucile Short. James Cotton Short, who was married to Bonnie Larue Short, received the Boyd-Short Farm when his father died. They were the parents of Virginia, Barbara, and Joan. An attorney, Short is still remembered as a General Sessions Judge, a position he held for 36 years, and as the City Judge of Fairview. Because of his judicial work, he limited farming to beef cattle and hay.

Tennie B. Boyd Short was the daughter of W.A. Boyd and grandmother of James Short.

Virginia Short Hunter, the 3rd-great-granddaughter of the founders, acquired the farm of just under 100 acres in 2005. She and her husband, Floyd Hunter, were the parents of Jeffrey, Allison, Christopher, and Farley. During their ownership, the main products were hay, wheat, soybeans, and cattle. Christopher Hunter acquired about 54 acres of the ancestral farm on the death of his parents. He and his wife, Vickie, are the parents of Dustin Short Hunter and Nicholas Alexander Hunter. The Hunters chose to remodel the twentieth-century house and restore the log cabin. They have 15 acres in hay and 22 acres in soybeans. The Short Farm was certified as a Century Farm in 2022.

SHORT FARM, 1807

Bonnie, Virginia, Joan, Barbara, and Judge James C. Short, the fifth owner of the farm. Daughter Virginia was the sixth owner.

The log house, restored, remodeled, and expanded over the years, was likely a slave dwelling. It became a home to a tenant family after the Civil War.

The house lived in by Judge Short and his family was remodeled by Chris and Vickie Hunter.

The multi-use barn remains a focal point of the farm's operations.

Chris Hunter is the 7th-generation owner of the farm, and received his 200-year Century Farm sign from Lynne Williams of the Tennessee Department of Agriculture at the Williamson County Fair event in 2022.

Grains stored in this silo were a primary crop grown for decades.

Sherwood Green, 1808

The Sherwood Green House remains a landmark on Rocky Fork Road in Nolensville. Surrounding the house is a log smokehouse and kitchen, a corn crib, the remnants of a slave cabin, and the family cemetery. (Photo by Skye Marthaler)

When traveling west on Rocky Fork Road in Nolensville, a marker placed by the Williamson County Historical Society tells the story of early settler Sherwood Green. Green, an active Methodist and Freemason, came into this area in the late 1700s from North Carolina as part of a team surveying Revolutionary War grants. The team was led by his father-in-law, Col. William Christmas, who is recognized as an early city planner and laid out the towns of Raleigh and Warrenton, North Carolina. Christmas became the Tennessee Surveyor General and Entry Taker in 1803. That same year, Green's wife, Martha, whom he married in 1800, and other members of the Christmas and Green family came to live in the Nolensville area, which was a wilderness. Green built a two-story log house on his 640-acre property and began clearing land to farm, while continuing his surveying profession. Martha died in 1828, having borne nine sons and two daughters. Sherwood Green's second wife was Mary Ozburn Johnson, daughter of a neighboring family. Edward J. Green, their only child, was born in 1830. Green, a slaveholder and planter, continued to acquire tracts of land in Williamson, Davidson, and Rutherford Counties. During his lifetime, Green gave each of his sons 750 acres and willed the homestead and house to his son Edward and his wife, Mary King, upon his death in 1840.

In this photograph from the early 1900s, the Sherwood Green House has evolved from a two-story log house into its present style, referred to as an I-house. It has a central passage and chimneys at either end. Those photographed are identified as the family of Lundy and Maude Green. Also included is a woman, holding the youngest child, who is likely part of a tenant or sharecropping family living and working on the property.

Edward and Mary Green were the parents of eight children. They lived in the log house, and by this time, there was also a log smokehouse, a separate kitchen, as well as slave quarters. A cemetery was set aside east of the house. It is likely that during their ownership, the house was expanded and covered with weatherboard. A second-story porch was added, and in the 1920s, a first-floor porch was added, giving the house its current appearance.

The Civil War saw activity near and around the farm from both Federal and Confederate troops. Edward and his older half-brother, Sherwood, Jr., fought for the Confederacy. Sherwood, Jr., and his neighbor and relative, John W. King, were both killed near Atlanta on August 24, 1864. For the remainder of the century, the Greens continued to engage in general farming and likely utilized tenants and sharecroppers, both black and white, to operate the extensive acreage.

Lundy Green, born in 1871 to Mary and Edward, became the owner of the house and farm in the 1920s. For more than two decades, he and his wife, Maude York, and their sons, Allen and John Edward, lived and worked on the farm. The dairy industry was a good option for farm families during this time due to the increasing demand for milk and milk products in growing towns. Lundy and his sons acquired an enviable herd of milk cows and operated what was long known as the Green Brothers Dairy.

Green Brothers was the name submitted by the family when the farm was first certified as a Century Farm. Allen Green, a U.S. Army veteran, and John Edward "John Ed" Green managed and operated the dairy farm for most of their lives. Allen married Bettie Battle, and they had Jennifer and Bill. John Edward "John Ed" Green and Ethel McMahon Green made their home on the farm.

In 1988, Janice and Ben Green, a nephew of John Ed and Allen, owned the property. Regent Homes purchased a portion of the land and developed Sherwood Green Estates, a residential community. Janice Green continues to live in and maintain the house, which has been featured in books and articles and is listed in the National Register of Historic Places.

Aurelia Acres, 1810

The family of William Shannon and Willie Demonbreun Patton are in front of their home in 1908. From left are Carl, Willie holding Mattie Reeves (Stewart), Douglas, Minnie Aurelia (the farm's namesake), Mary Cleo (Roberts), and William.

Just a few years ago, Patton Road was a scenic lane of working farms connecting Cox Road and the Horton Highway. With the opening of sections of I-840 between 2000 and 2002, development quickly followed. Arrington Vineyards is a well-known enterprise and destination at the west end of Patton Road. Traveling from either end along the route, the remaining fields, fences, livestock, barns, and historic homes that still stand mark this as a long-time farming community. The Patton family and their descendants have lived here, watched the changes, and adapted while maintaining a farm founded when Williamson County had been in existence for little more than a decade.

Jason Patton acquired three hundred acres on Nelson Creek in 1810. The following year, he married Bethenia Bostick, who lived on a sizeable farm between Triune and Arrington. Her parents were Mary Jarvis and John Bostick, a Revolutionary War veteran and planter in Williamson County (this pioneering couple is buried in a gravesite in the King's Chapel development on Highway 96 on land they once owned). The Pattons grew corn and tobacco, as well as vegetables and fruits, allowing them to be self-sufficient. The mules and horses they raised were necessary for transportation, farm work, and income.

William Douglas Patton, son of the founders, was the next owner of the property. He married Mary Jane Patton, a first cousin, and they raised livestock and crops. The two-story frame house was likely built during this couple's ownership.

Likely built by William Shannon and Willie Patton circa 1890, this was a popular style of the time referred to as an up-right and wing house. It remains the residence of their descendants.

Their son, William Shannon Patton, acquired the farm around 1880. He and his wife, Willie A. Demonbreun, had seven children. Patton was known for his horse breeding and won numerous awards. The three daughters of William and Willie — Mary, Minnie Aurelia, and Mattie — were the fourth-generation owners in 1933. Significantly, Patton Road became a county road, and a bridge was built to replace the ford over Nelson Creek in 1935. During this time of change, electricity was extended to the farm, a kitchen was attached, a bathroom was installed, and running water was piped to the house. Barns for tobacco, cattle, and hay were built to accommodate increased production.

Mark Stewart, son of Mattie Patton Stewart, acquired sixty acres of the original farm in 1997. He and his wife, Carolyn, are the parents of Marcus and Scott. Carolyn was an educator, and she and Mark enjoyed hosting friends and relatives at the farm. During their ownership, the owners made improvements and upgrades to the two-story frame house.

Mark Stewart, the owner of Aurelia Acres, weaves a tall tale about mules at the annual Century Farm dinner at the Williamson County Fair.

Mark is an ordained United Methodist Deacon and has served in ministry for 35 years in various positions, continuing in volunteer ministry for 25 years. He also taught math and music in junior high school and was stationed in Germany while serving in the U.S. Army. Mark is an award-winning teller of tall tales, regaling audiences with humor, wit, and a delivery that leaves them laughing. After Carolyn died in 2010, Mark remained on the farm, and he and Scott share the house when Scott is not traveling for his work as a musician. Scott hopes to continue ownership of the family farm for another generation. The land is leased to nearby Devlin Farms, where a variety of certified organic vegetables are grown.

Mark describes some of the changes he has witnessed, including the upgrade of the road from dirt to crushed rock and then to the smooth paved road of today. He notes that subdivisions, new homes, and businesses are all around, and the traffic on Patton Road, including both commercial and individual vehicles, increases each year. Mark easily talks of the farm where he has lived for the past twenty-five years and where he visited relatives for the previous sixty-three years. One of those relatives was his Aunt Aurelia, for whom he named the farm when it was certified as a Century Farm in 2017. Mark reminisces that he and his wife enjoyed sitting on the front porch, watching wild turkeys and deer, and feeling blessed to live on the farm. "We decided," he recalls, "that this is as close to heaven as you can get on this side." As if giving a benediction, Mark writes, "I thank those who came here in 1810 and those who followed for blessing me with this farm."

AURELIA ACRES, 1810

The scenic country lane that is Patton Road remains mostly intact — for now.

From Patton Road, the late 19th-century farmhouse can be glimpsed through the trees.

The fences, barns, and pastures of Aurelia Acres show a productive and well-maintained farm. Organic vegetables are the primary crop.

WILLIAMSON COUNTY BECOMES A LEADER IN AGRICULTURE 1815-1859

To analyze and understand this period in American and Tennessee history, volumes have been written and will continue to appear as more research is discovered, gathered, interpreted, and reinterpreted. The years following early settlement and before the Civil War are referred to in different ways, depending on the author and the purpose of the writing. All concluded that this time was complex, and most issues in politics, religion, the economy, culture, and society were directly or indirectly associated with agriculture. The Century Farm stories provide an avenue to bring these years and the people, both those who owned land and those enslaved, into focus.

Williamson County, though just established in 1799, quickly became an agricultural leader in Tennessee, ranking either second or third in the state in the decades leading up to the Civil War. Farms were generally productive due to the quality of the soil, the availability of water, and the climate. The records of Century Farms of varying sizes describe the variety of livestock and crops in the different sections of the county. Timber was a basic and lucrative commodity, and sawmills operated in several communities. Corn was the staple grain with wheat and oats, and some rye, yielding significant amounts annually. Grist mills were scattered across the county to accommodate farmers, and surplus grains were marketed to other areas by water, wagon, and, after 1855, rail. Horses and more often mules, beef and dairy cattle, swine, and sheep were raised on large and small holdings. Therefore, hay was a necessary and primary commodity. Cotton and tobacco were the leading row crops in terms of yield. Subsistence farmers might have one mule for field work and transportation, a milk cow, a vegetable garden, and a cash crop, such as cotton, tobacco, or grain. Agricultural fairs emerged during the 1850s to showcase farms and farming practices. Williamson County's first documented fair was in 1857.

As property owners prospered, the houses in which they lived evolved from log to more substantial dwellings of stone, brick, and frame. Some families chose to incorporate the original log house, add rooms, and remodel it over time. This trend is illustrated in the architectural history of several Century Farms, as houses, barns, and outbuildings of log were retained. At other farms, buildings originally built by the first settlers were torn down and their materials repurposed. Brick and frame houses became more available and affordable to accommodate expanding families. Changes in style, materials, and construction techniques advanced, and dwellings were furnished to reflect the owner's taste and economic status. Examples of several architectural styles prevalent in Tennessee and other regions also appeared in Williamson County, including the "I-House," commonly associated with the antebellum period and built in both frame and brick.

An "I-House" with chimneys at either end, a symmetrical façade, a two-story portico, and a rear ell was a popular style in brick or wood that appeared across Williamson County and the South in the antebellum period and into the twentieth century. This one was built in 1855. In contrast is a typical still-extant single-pen slave dwelling from 1830, featuring stairs to a loft and a stone chimney. After 1865, it was a tenant house for several years. Both buildings are equally significant properties in Williamson County and each is listed in the National Register of Historic Places.

Slavery existed in Williamson County from the time it was settled until the end of the American Civil War in 1865. Not all slaves were owned by planters and farmers of extensive holdings. Records note that small farms often had one, two, or more men, women, and children. Merchants also relied on slave labor. In 1820, the slave population in Williamson County numbered 6,792, while the white population was 13,858. By the 1850 census, people held in bondage numbered 12,467, with whites counted at 14,337. By 1860, the number of slaves outnumbered that of whites, at 12,152 to 11,315. Because of the work and knowledge of slaves in all aspects of agriculture, and their skills and experience in animal husbandry, knowledge of crops, blacksmithing, and construction, as examples, the advancement of the county and the economic status of their owners was due in large part to their labor.

Those enslaved, however, profited little from the prosperity that continued through the first six decades of the nineteenth century. Their housing was primitive and remained largely unchanged as the years progressed. Most lived in a one-room log or frame dwelling with a chimney and perhaps a sleeping loft. The individual or family might be accorded their own vegetable patch. Slaves with special skills, such as seamstresses, woodworkers, or horse and mule trainers, were sometimes allowed to work for others and might be permitted to keep a portion of their pay. They were, however, still slaves, and basic freedoms were denied.

As farms expanded and improved, more and larger outbuildings were built specifically for livestock and for grain storage. Corn cribs and smokehouses for curing meat, primarily pork, and chicken houses were among the first dependencies to appear on a farm, and they were enlarged or rebuilt as needed. Separate kitchens were common because of the heat and the danger of fire. Many families, though, cooked over an open fire either outside or in the fireplace of the dwelling. Spring houses were constructed not only to keep the water supply clean and readily accessible but to keep butter, milk, and fresh or canned foods cool. Fruit trees and vegetable gardens were a necessity on farms, and town lots often supported chickens, a milk cow, a horse, and substantial kitchen gardens.

Franklin's businesses became more numerous and diverse with each decade, and other towns and villages grew, including Poplar Grove (later College Grove), Fairview, Brentwood, Nolensville, Thompson Station, and Benton Town (later Hillsboro and Leiper's Fork). Communities throughout the county identified themselves with a general store, livery and blacksmith shop, churches, a cemetery, a school or academy, and, often, a post office. Williamson County had changed immeasurably, but its economy, society, and culture were largely dominated by agriculture.

WILLIAMSON COUNTY BECOMES A LEADER IN AGRICULTURE 1815-1859

POYNOR FARM, 1815

WOODLAND VIEW, 1820

BLUE GRASS, 1825

VALLEY VIEW, 1827

CANNON FARM, 1842

HATCHER FAMILY DAIRY, 1847

BAG END, 1848

PLEASANT VIEW, 1848

REAMS-JEFFERSON AND JEFFERSON FARMS, 1854

THE WILLIAM STEELE FARM 1859

Poynor Farm, 1815

The W. S. and Sophronia Peach family of Peach Hollow ran the store in this 1910 photograph. It was one of numerous small country stores in Williamson County that were so important to their communities.

Peach Hollow, off Garrison Road and south of Leiper's Fork, is near the Natchez Trace Parkway and I-840. When the Poynor Farm was founded over two hundred years ago, however, this valley was a lonely wilderness. The history of this farm illustrates how neighboring families intermarried over the years, and generations of ownership did not always proceed from parents to children, but to other relatives. Cousins, nephews, nieces, and other relations were often heirs and in land transactions, making the task of tracing the owners of a specific property complicated. Claudine Poyner spent hours patiently researching deeds, census records, and family papers to prove the farm had been within the family for more than two centuries. The Poynor Farm was first certified as a 100-year farm, but subsequent research traced its history back to 1815, as her diligent investigation into the complex lineage proved. Other Century Farms have been able, through additional research, to prove earlier dates of founding than those indicated in their first application.

Four generations were on hand to receive the 200-Year Century Farm sign: Douglas and Claudine Poynor; Irene Yates Poynor, holding the sign; and Robbie, Tracy, Haley, and Will Poynor.

It was not until the seventh owners of the farm, William Thomas Poynor and his wife, Virginia Burns Poynor, that the Poynor name was associated with the property. Before that time, relatives with the surname Burns and Peach lived on and worked the land. All, however, trace their line to George Burns, who bought 87 acres from John Campbell in 1815. George Burns, father to William and James Burns, purchased portions of the farm at different times. In the 1850 census, crops and livestock grown by James, his wife, Flora Church Burns, and their children were sweet potatoes, corn, oats, cattle, sheep, and swine. The products were butter, cheese, and wool.

By the 1880s, William Strickland Peach and his first wife, Sophronia Ann Burns Peach, the niece of William Burns, were the owners of 190 plus acres of the farm. They were also third cousins. William was a Methodist Circuit Rider for fifty years, preaching at Garrison, Theta, and Sycamore churches. He also ran stores at Garrison and Sycamore. Census records for this family note that they had one horse, swine, chickens, and a sorghum mill. Some longtime residents know the store at Garrison was run by family members, Virginia Burns Poynor, and later by her daughter, Annie Pearl.

Irene Yates Poynor, the grandmother of the current owners, and her husband, William Ewen Poynor, acquired acreage previously owned by his grandparents. They and their children, Douglas, Brenda, and Keith, raised cows, sheep, horses, swine, and a donkey. The children participated in 4-H projects and raised cows to show. Douglas married Claudine Poynor, and their children, Robert "Robbie" and Chris, were active in 4-H, showing sheep.

Robbie, Haley, Will, and Tracy take a rare break. Farm work, especially with animals, goes on even at Christmas.

Robbie and Tracy Poynor are the current and ninth owners and operators of the farm. A multi-generational farmstead, they, along with their children, Will and Haley, live on the property, as do his parents, Claudine and Douglas Poynor. Registered shorthorn cattle, registered Boer goats, honeybees, and wildflowers are included in the family's award-winning livestock and products. Tracy and Robbie's Honey was awarded the Grand and Reserve Champion ribbons at the Williamson County Fair in 2018. Among the many accolades are Williamson County Grand and Reserve Champion Wether, 2024, 2023, 2021, 2020; Tennessee State Reserve Champion Shorthorn Steer, 2024; Tennessee State Reserve Champion Main-Anjou Steer, 2025. A true working family farm, all the generations associated with this place have contributed to the farming traditions in and beyond Peach Hollow since the early days of settlement.

Working at the sorghum mill on the farm were W. T., Dewey, Jerry Poynor, and a team of mules.

The Poynor Farm is renowned as a thriving livestock farm, and its animals consistently win in several categories of the Williamson County State Fair.

The Poynor family is the recipient of awards for livestock and honey. One of the biggest wins was being awarded the Grand Champion and Grand Champion "Farm Bred" goat for Tennessee in 2024.

Woodland View, 1820

Woody, Tom, and Thomas Herbert, along with the entire family, are proud of the 200-year-old Century Farm sign displayed on their ancestral farm.

Few farms exist in the highly developed commercial and residential northern part of Williamson County, though that area was traditionally one of the richest agricultural regions. With a Brentwood address on Old Smyrna Road, Woodland View is a surviving anchor and reminder of the prosperous farms that once filled the landscape. Brothers Tom and Bob Herbert run beef cattle and grow hay on 55 acres of an initial 100-acre farm. They trace their family to Richard Herbert, who immigrated from England and purchased land in 1820. Records indicate he also had cattle, along with swine, grains, and most certainly, vegetables and fruits. A log barn built about 1845 by Richard's son, Robert N. Herbert, still stands within a newer barn. A spring house also dates from the first two generations, as they and their families worked to build and improve the farmstead and to expand it to 213 acres in the 1890s.

"Roose" Herbert, an industrious and well-respected farmer, was very active in the community despite being wheelchair-bound for most of his life.

George O. "Buck" Herbert, Sr., the founder's great-grandson, acquired the farm in 1918 at the age of 14. He and his mother, his wife Gladys, and his son, R.N. Herbert, II, lived there, raising livestock and growing hay and grains. George had a route selling milk and eggs in Nashville that were produced on the farm. R.N. "Roose" Herbert was a star football player for Nashville Central High School, but at a practice in 1950, he was injured, resulting in paralysis. Though in a wheelchair for the rest of his long life, he did not allow that physical challenge to keep him from being a successful and active farmer. "That's what kept him going," said nephew Tom Herbert on his uncle's death in 2011. "He did all he could do, and more than most walking people." He drove a tractor to bush hog and bale hay. He also had a specially equipped van that allowed him to herd his cattle, which followed his call to them. Roose and his family also hosted his annual "Surprise" birthday party for the Nashville Central alumni he played with. Neighbors and others also came to learn from him and to honor their friend.

A large parcel was sold by the fifth generation for a subdivision, Whetstone, in 2005. Tom, a realtor, and Bob Herbert, the sixth-generation nephews of Roose, acquired the farm and continue to live on and manage the remaining acreage. The Herberts run cattle and grow hay. The 1845 barn and its later outbuildings continue to play a role in the farm's daily operations, which is now over 200 years old.

> Of Woodland View and farming, Tom Herbert comments, "If we had not inherited the farm from our ancestors, we would not be farming here now. UT Extension in Williamson County, TAEP (Tennessee Agricultural Enhancement Program), and TCA (Tennessee Cattlemen's Association) organizations have helped us and other farmers stay in business over the past few years." He reflects on daily challenges, stating, "As you know, farming in Williamson County and in the city limits is becoming more difficult. It's like we are on an island with city folks all around that don't understand farm life." He cites this specific example, "people have cut my fence thinking they can utilize my property as theirs and then call the police when cows get in their yard." Farmers whose roots run deep in Williamson County, who add immeasurably to the economy, the food supply, the scenic beauty of the area, and who take care of the land as their family has for generations, deserve better than they often receive from those living nearby who show little regard for agriculture and the professionals who practice it.

The children of Robert Nathaniel and Elizabeth Lewis Cummins Herbert: James Harvey, Mary Selina, John Overton, Thomas L., George Washington Sneed, David Cummins, and Robert N. Herbert, Jr.

Within a much-later mule barn is a two-story log crib built circa 1845.

The hay barn and squeeze chute are an integral part of the cattle business.

The 8th generation of Herberts feeds the herd.

Baling hay in the shadow of development.

Tom and Bob Herbert are the current owners.

Blue Grass Farm, 1825

Within the main residence is the log cabin that became the family dwelling after a larger house and outbuildings were destroyed during the Civil War.

John B. Bond is one of the first members of his family to come into what would be southeastern Williamson County. Born in Ireland in 1760, John traveled in the Northwest Territory and through what became, in 1803, the state of Ohio. Here he met Elizabeth Bryan and, as family history recounts, she left her home, taking nothing but her Bible, which remains with the family. They traveled south to Tennessee and the Bethesda community, where Revolutionary War soldiers were settling land grants, the genesis of several family farms that exist today.

Records indicate Bond operated a blacksmith shop as early as 1797, two years before Williamson County was established. While some farms may have been established earlier than the date of certification, obtaining proof from these early years can often be a challenge. Records, however, indicate that John and Elizabeth Bond and their family were farming this land in 1825, on 715 acres. The family, which eventually included ten children, grew tobacco, sheep, and subsistence crops.

Cicero Columbus Bond with his son, James William Bond.

Cicero Columbus "C.C." Bond inherited the farm in 1848. A magistrate representing the 12th District, C.C. managed a diverse farm of tobacco, corn, wheat, and livestock. As with most of the county's farms during the Civil War, soldiers from both sides foraged for food, took livestock, tore down fences, and stripped buildings of wood for fires. Cicero's farm was in such a state of ruin that he had little option but to move to a one-room log cabin and begin the process of rebuilding. For the remainder of the century, he and Rachel Blythe Bond, along with their children, worked to bring the farm back into a profitable operation. Their log house is the nucleus of the farmhouse that remains the primary dwelling today.

James William Bond acquired the family farm in 1909. James was the great-grandson of the founders and received title to the family landholdings in 1953. Sixteen years later, brothers Charles and Dan Bond acquired 420 acres of the original farm. For several years, they managed 870 acres and operated a Grade A dairy. They also raised tobacco and beef cattle. In 1984, the Bonds "received a certificate from the American Polled Hereford Association for 64 years of continual breeding of Polled Hereford cattle." Dan and his family now live on and operate Bud's Long View Century Farm, founded in 1900. Charles and his wife, Carol, and their sons, Allen and Robert, continue to raise beef cattle and hay, and they have acreage in pastureland. Charles and Carol also own the Bond Farm, which was established in 1870.

Today, Blue Grass Farm remains a major agricultural operation primarily producing beef cattle and hay.

Valley View Farm, 1827

The original log dwelling within the current house is much more comfortable today than when built circa 1827.

In the southeastern part of the county, Kerry and Sharon Connell have restored and live in a house that encompasses the circa 1827 log house built by Allen F. Wood, who founded Valley View Farm. Wood and his wife, Sarah, raised livestock and row crops on 50 acres, which he expanded to 147 as he established his farm in the Flat Creek community. In 1881, the Woods gave their daughter, Mary Ann Wood Sanford, 50 acres of the farm. Married to Archer Wood Sanford in 1850, she gave birth to four children, including Stephen, Sallie, and Minor. The family raised a variety of crops and livestock for themselves and for sale. It was likely during this time that the house evolved from a log dwelling to the gable-front and wing design. A faded photograph of Sallie Sanford, who died in 1921, shows the current configuration of the house. Minor Sanford was the next owner of the farm. He married Era Jane Mosley in 1901. They were the parents of Robert and Albert Sanford.

VALLEY VIEW FARM, 1827

Mary A. Wood Sanford and her daughter, Sallie Sanford, were the second and third generations. They share a grave marker in the Lester Cemetery. Photograph by John C. Jones

Robert Sanford, the great-grandson of the founders, acquired 147.5 acres of the farm in 1945, and he and his wife, Viola Ward Sanford, worked and lived on the farm for at least 40 years; Robert died in 1986, and Viola in 2001. During their ownership, the property was expanded to over 200 acres, and they specialized in beef cattle and hay. Valley View was recognized as a Century Farm in 1986. Their daughters, Era Ann Sanford White and Mary Margaret Sanford Connell, grew up on the farm. Mary Margaret married Terry L. Connell in 1947.

Valley View, home to Sharon and Kerry Connell, is a well-known landmark on Flat Creek Road. They continue the agricultural traditions of the Wood-Sanford families while appreciating and maintaining the historic landscape, outbuildings, and house.

Livestock and row crops are staples of Valley View Farm, which continues to utilize outbuildings constructed over the years.

The current generation gathered at the farmhouse on a recent Easter.

The farmhouse is truly festive at Christmas and has hosted many special occasions and holidays of the Wood-Sanford-Connell families.

Cannon Farm, 1842

Col. Hardy Murfree, a Revolutionary War veteran for whom Murfreesboro in adjacent Rutherford County is named, is buried on the Cannon Farm.

Adjacent to eastern Williamson County is Rutherford County, which was created from parts of Williamson, Davidson, and Wilson Counties in 1803. In 1811, Murfreesboro became the county seat and both were named for Col. Hardy Murfree, a Revolutionary War veteran who was born in Murfreesboro, North Carolina. That town was named after his father, William Murfree. Through land grants and purchases, Hardy Murfree acquired thousands of acres in Tennessee, some in what is now Williamson County. When Col. Murfree died in 1802, he was buried in his family cemetery on what is now the Cannon Century Farm, founded in 1842.

The Cannon Farm is associated not only with the pioneering Murfree family but with other early settlers and names familiar in county and state history. The Perkins family owned hundreds of acres and built some of the finest and still extant antebellum homes in both rural parts of the county and Franklin. In 1842, Samuel and Nancy Perkins bought large tracts that were part of the lands granted to Murfree for his service in the Revolutionary War. The Perkins, who had five children, primarily raised swine, cattle, and sheep, as well as crops such as cotton and corn. The couple left each of their children a substantial farm.

Their daughter, Susan Agatha Perkins, and her husband, William Perkins Cannon, received 775 acres east of Carter's Creek Pike extending from West Harpeth Road to Thompson Station Road. William, born near Triune, was a veteran of the Seminole War and the son of Newton Cannon, eighth governor of Tennessee. Gov. Cannon fought in the Creek War and was a leader of the Whig party who regularly disagreed with Andrew Jackson. He also served in the United States House of Representatives. William's sister was Rachel Adeline Cannon Maney, mistress of Oaklands Mansion and plantation in Murfreesboro, Tennessee, now a historic house museum dating from circa 1818.

William Perkins Cannon married Susan Agatha Perkins, who received 775 acres from her father. They managed the farm before, during, and after the Civil War.

Susan and William Cannon were major landowners and planters before the Civil War and depended on slave labor to harvest row crops and tend to livestock herds. The farm was significantly damaged when Federal troops camped there during the Civil War. Their son, Newton, returned from the war to help repair and rebuild the farm. He also wrote *Reminiscences of Newton Cannon, First Sergeant in 11th Cavalry, C.S.A.*, from his perspective as a teenager serving with Nathan Bedford Forrest and Joe Wheeler. Of his children with wife Virginia Brown McEwen, one son served in the Spanish-American War and one in World War I, thus continuing a tradition of soldiering that began with the Revolutionary War.

Edgar Brown Cannon and his wife, Marguerite, lived on and managed the farm for much of the twentieth century. The Cannons also added 90 acres to the original acreage, and along with their son Ed, lived in the mid-nineteenth-century dwelling. In the 1950s, Cannon had local builder S. E. Farnsworth construct a tongue-and-groove barn that was intended for mules and hay. It is still in good repair today and is used for cattle management, as well as for storing hay and equipment. Tobacco was a major crop in the twentieth century, and two barns built specifically for that crop remain intact. A springhouse and blacksmith shop are relics of a time when the family was primarily self-reliant. For the past several years, neighboring farmers, the Saunders, have raised cattle and hay on the Cannon Farm.

Edgar Brown Cannon

Wheat thrashing with a steam engine and mule-pulled wagon as machinery transitioned from animals to other sources of power.

The mule barn, built by Farnsworth, remains in good repair, though no mules are on the farm today.

The Cannon Farm is one of the most productive and picturesque.

Hatcher Family Dairy, 1847

Accepting the 150-year Century Farm sign at the 2024 Williamson County Fair are Sharon and Dr. Charlie Hatcher, Tennessee Commissioner of Agriculture. Representing two more generations of the family are Mary Morgan Gentry, married to Charles Hatcher, and their son, Cannon.

The 2024 *Williamson County Historical Journal* includes the article, "Hatcher Family Farm: Abram W. Hatcher's Family Legacy," by Mary Morgan Gentry. Growing up on Pleasant View Century Farm, also known as the Gentry Farm, Ms. Gentry is aware of the significance of working family farms to the heritage, landscape, environment, and economy of the county and state. After marrying Charles Hatcher in 2019, she researched deeds at the Williamson County Archives, family records, and interviewed relatives to learn more about the Hatcher family's history. The product is a well-documented Century Farm application that serves as the basis for her article.

William and Lucy Hatcher moved from Bedford, Virginia, to Tennessee in 1819. Their eldest son, Octavius Claiborne "O.C." Hatcher, who married Caledonia Pillow, purchased a 270-acre farm in the College Grove area in 1847. They had five children and raised livestock and grains. O. C. Hatcher, a veteran of the Mexican War, died of typhoid fever when he was forty. His will directed that the farm be sold at public auction. His brothers, Thomas Logwood Hatcher and Spotswood Henry Hatcher, purchased 265 acres in 1860. Caledonia moved to Lincoln County, where four of her children lived in or near Fayetteville.

Abram Wooldridge Hatcher, O.C.'s younger brother, married Mary Susan Dodson in 1858. In the 1860 census, he and his wife, along with their first son, were living on the 266 acres owned by his brothers. Abram enlisted at Camp Trousdale in December of 1861 and was away from the farm much of the war. He was, however, on leave in November of 1864 when the Battle of Franklin was fought only a few miles away. The family could hear the cannon from their yard. In August of 1865, Mary died from complications of childbirth with their third son, who also died a month after his mother.

Abram and Martha Chreisman Hatcher were married in 1868, the same year he became the owner of the farm.

In 1868, Abram was officially deeded the property by his brothers. In that same year, Abram married Martha E. Chreisman of nearby Bethesda, and they were the parents of eight children. They likely built the house that stands today in the 1870s. Family records indicate that Martha churned butter and sold it locally along with eggs, honey, and meat. Row crops included millet, rye, wheat, sorghum, clover, and corn. In addition to the Jersey dairy herd, the family raised and sold cattle, hogs, sheep, and turkeys.

The youngest son of Abram and Martha was George Hatcher, who was just sixteen when his father died. He and his sisters lived with their mother, and together they worked on the farm. George attended the People's School, later Battle Ground Academy, working as a janitor to pay his tuition. He studied medicine at Vanderbilt University and married Eula Neely in 1912. Later that year, following his mother's death, he purchased 145 acres from her estate. His sister, Lucy, and her husband, Rucker Miller, bought the homeplace and 101 acres.

Lucy Hatcher and her husband, Rucker Miller, lived in the Hatcher homeplace, built by her parents circa 1870.

George Abram "Abe" Hatcher was a WWI and WWII veteran.

Dr. George Hatcher served in World War I and World War II, practicing at Central State Hospital in Nashville and later at Overbrook Hospital in New Jersey. He was also an attorney and a progressive farmer who instilled in his children a sense of discipline, work ethic, and appreciation for the family farm. He and his family visited the farm often and had a small log cabin moved from Bethesda and reconstructed on a site near a spring on the property. That cabin still stands today and is easily recognized as part of the farmstead. During World War II, Eula moved to the cabin and ran the farm herself because her son, Abram "Abe," and her husband were serving their country. In 1945, when George returned home, they purchased the home place and land that belonged to Lucy, who had died during the war. Dr. Hatcher had Guernsey cows and built the present-day milking barn and a tobacco barn. In addition to the dairy cows and tobacco, the Hatchers had hay, corn, hogs, sheep, and goats.

After serving in Japan and the Philippines during World War II, Abe attended the University of Tennessee, earning degrees in Chemistry and Agronomy. Later, he attended Cumberland University Law School and met Jacqueline Price, who was studying music. They married in 1953 and moved to the home place on Arno Road. His patent law career took him to many places, and his and Jackie's five children were born wherever he was working at the time. In 1972, however, they returned to the farm, though he continued to practice law in Franklin. Before his death, he gave each of his children a portion of the family farm. Four of the five siblings still live on the farm, making it a true family farm to this day.

It was an exciting day when the family gathered for the last hand milking before the robotic milking system was installed in March 2024. From left: Mary Morgan Gentry, Charles Hatcher, Dr. Jennifer Hatcher, Sharon Hatcher, and Dr. Charlie Hatcher. In front, the next generation: Cannon Hatcher and Hatcher Yoest.

Dr. Charles Hatcher, son of Abe and Jacqueline, was given 145 acres of the farm in 1978. He and his wife, Sharon, are the parents of Charles and Jennifer. Dr. Hatcher and his brother Jim were managing the farm and milking cows when they came to what the family refers to as "a crossroads" in 2009. They could either sell the herd and end the long tradition of dairy farming, or begin bottling their own milk and selling it directly to customers. Fortunately for them, as well as area residents and vendors, they chose the latter course of action. In 2009, the Abe Hatcher Creamery was built on the farm and now sells to over one hundred outlets in Middle Tennessee. Touring the dairy, hearing the stories of the generations of Hatchers, and seeing where milk really comes from are popular with school groups and visitors of all ages. Dr. Charlie Hatcher served as the Commissioner of Agriculture for the State of Tennessee for seven years until September 2025. Dr. Jennifer Hatcher, a veterinarian like her father, runs the clinic and lives on the farm with her husband Chuck Yoest and their two children, Hatcher and Charlee. Sharon is the calf and chicken manager, bookkeeper, and maintains the two Airbnbs on the farm. Charles is president of Hatcher Family Dairy, which recently installed robots to milk the cows. Mary Morgan and Charles are the parents of Cannon and Martha Jane, who, along with their cousins, are the next generation to learn first-hand the business and the agricultural traditions of the Hatcher Family Dairy.

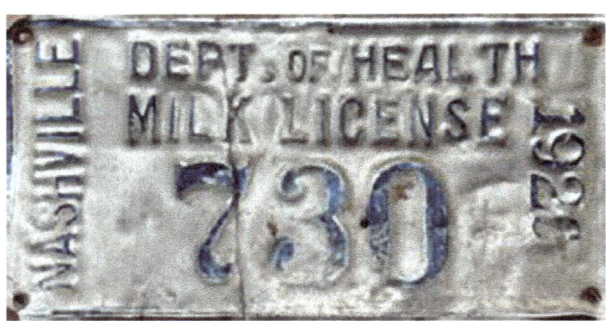

The first Hatcher Milk License plate from 1926 was found in the rafters of the original milk barn.

Twice a day, the milking barn is the busiest building on the farm.

Calves and cows are the much-photographed stars on the tours of the Hatcher Family Dairy complex.

Bag End, 1848

The historic Smithson-McCall-Fisher house incorporates the original 19th-century log dwelling.

Farm owner and librarian, Susan McCall Fisher, named her ancestral farm after the safe, comfortable, and charming dwelling of Bilbo and Frodo Baggins, whose adventures are chronicled in *The Hobbit* and *The Lord of the Rings* by British author J. R. R. Tolkien. In fact, Bag End Farm has remained a comfortable and charming home through several generations, and it is safe from commercial development. That is due to the foresight and planning of Susan and her husband Steve Fisher, who, in 2002, placed the 243 acres in a conservation easement with the Land Trust for Tennessee. The easement, which stays with the property, even if it were to pass out of the Fisher family, also limits residential construction to two additional homes. The Fishers have a unique perspective on the past and the present of the community, for Steve was the principal of Bethesda Elementary School and Susan managed the Bethesda Branch of the county's library system. They hope the unbridled growth seen in nearby Spring Hill and Thompson's Station does not extend into Bethesda for some time, but, if so, they have preserved their farm.

Matriarch and farm founder Nancy Smithson's grave marker in Giles Cemetery is still legible. (Photograph by Bob Barnhill)

Flat Creek is a historic community established in 1799 in the southeasternmost corner of Williamson County, Tennessee. It was founded primarily by Revolutionary War veterans who were awarded claims for their service. Isaac Gillespie, Thomas Gillespie, Jr., and David Gillespie inherited land in Flat Creek from their father, and they are buried at the Moses Steele Cemetery near Flat Creek. Bag End, sometimes referred to as the Smithson–McCall Farm, was built on 528 acres of Thomas Gillespie's 4,000 acres that were willed to his son David and conveyed to him in 1808. Samuel Henderson purchased 125 acres from David in 1813, and it was from him that Nancy Pettus Smithson, widow of Clement Smithson, bought 125 acres in 1848. It was unusual for a woman to own property during this period because laws and society did not favor female ownership. She lived with several of her children in a log house likely built by Henderson. Her son, Charles E. Smithson, became the owner of the farm in 1855. He and his wife, Jane Giles Smithson, were the parents of nine children. One of their sons, Charles T. Smithson, a Civil War veteran, became the owner of the land in 1891 that included the two-room dog-trot cabin and log kitchen.

Alice Smithson McCall, daughter of Charles T. and Martha Smithson, was the next generation to own the farm. She and her husband, Andrew Lycurgus McCall, managed a diverse operation that included beef and dairy cattle, sheep, tobacco, and wheat. Their son Herbert Lycurgus McCall was born in the family dwelling and worked with his parents. He attended Bethesda School and Battle Ground Academy, and, after marrying Mildred Creswell, they lived on the farm. Herbert taught school, was the principal at Flat Creek, was a Mason and Shriner, an active member of the Farm Bureau, and a Sunday School teacher at Bethesda Methodist Church. Mildred was known in the community as a wonderful homemaker and for her large flock of Rhode Island Red chickens. Their son was Gerald McCall, who was born on the farm but chose another career, attended West Point, and reached the rank of full Colonel in the United States Corps of Engineers. He was, however, buried on the family farm in 1999, by which time his cousin, Susan McCall Fisher, owned the property.

Some of the land is currently leased for cattle, and a section is planted with native grasses. A vegetable garden and fruit trees add to the food selections at this peaceful place. The original log house, now part of a white frame house, and the original log smokehouse are reminders of the farm's long history. Susan and Steve raised their family on the farm, hoping their children and grandchildren would care for and return to this land as past generations have for decades. In an interview with Carole Robinson for the *Williamson Herald* (November 29, 2012), Susan explained, "We love the farm and wanted to protect it – for it not to be developed." She continued, "I grew up visiting the farm – I love it still, and I am happy knowing it will be protected in perpetuity."

The log smokehouse at Bag End dates from the first half of the nineteenth century and was on site, along with a log cabin, when Nancy Smithson purchased the farm.

The fields of Bag End Farm are protected from development by a conservation easement with the Land Trust for Tennessee.

Pleasant View Farm, 1848

Pumpkin season at Gentry's Farm is a highly anticipated event among locals.

Popularly known as the Gentry Farm among families who enjoy the corn maze, pumpkins, and hayrides each Autumn and by school groups that visit seasonally, this is a working farm that also teaches folks of all ages about farming—then and now. That was an intentional decision made by current owners Allen and Cindy Gentry, who hosted their first school group in 1989, then opened their farm to Fall and Spring field trips and visitors. This agritourism venture, one of the first in the state, is a way to support the farm family, protect the land, and educate a population removed from farming and any understanding of agriculture's importance to our daily lives.

PLEASANT VIEW FARM, 1848

Samuel F. Glass, Jr. and his wife Agnes Hunter Glass completed the two-story brick house following the Civil War. The dwelling was added to the National Register of Historic Places in 1988.

Samuel F. Glass, Jr. and Agnes Hunter Glass

Pleasant View, three miles west of Franklin, was founded in 1848 by Virginians Samuel F. and Sarah Malone Glass when they purchased 449 acres from Meredith P. Gentry. Primarily a cotton plantation at that time, Samuel managed the farm. His father moved to Franklin in 1812 and owned and operated a hat factory just off the square. Samuel began work on the substantial two-story brick house, but died in 1859. After the Civil War, Samuel F. Glass, Jr. completed the dwelling where he and his wife, Agnes Hunter Glass, lived and raised their four children. The farm's products included cotton, dairy cattle, and swine. A successful farmer and businessman, Glass, unlike many neighbors who slowly, if ever, recovered from the Civil War, had secured $100,000 by 1873, enabling him to purchase property once owned by the Perkins, McGavock, and DeGrafenried families.

Corinne Glass Gordon and husband Edward Allen Gordon were the third-generation owners and continued raising tobacco, hay, and livestock. By this time, cotton had ceased to be "King," and tobacco had become the primary cash crop on many farms. The family ownership continued with Corinne Gordon, named after her mother, and her husband, Hugh Channell. The family moved to Franklin around 1900, and the house was rented to tenant farmers.

Rebecca Channell Gentry, daughter of Corinne and Hugh, became the owner of nearly 500 acres of her family's farm in 1974 and returned to the farm in 1975. Married to Jimmy Gentry for 59 years, they were the parents of Allen, Scott, and James, Jr. Jimmy started a summer day camp around 1970 with some of his BGA students whose mothers wanted their sons to "work" on the farm. The camp continues today as an educational opportunity for local children. Allen started raising cattle in 1972, paying his way through MTSU, and then began managing the farm.

Allen and Cindy Gentry, along with their children and grandchildren, continue the legacy of Pleasant View, also known as the Gentry Farm.

When Allen married Cindy White in 1983, the farm began to focus on new opportunities that always included the traditions, buildings, and stories of past generations. Their teamwork has brought stability to the farm in the twenty-first century with hard work and creativity.

PLEASANT VIEW FARM, 1848

Before she and Allen married, Cindy, with a degree in Historic Preservation from MTSU, taught the Heritage Foundation's first History Classroom. She designed and presented activities that taught schoolchildren about the county's history, including farming, geology, and architecture.

After marrying Allen, the couple moved to the farm, where they combined a log cabin and a former tenant house to create their unique, comfortable home. Located just behind the main house, they raised their three children, Mary Morgan, Hope, and Jase, in the home that could be said to symbolize their dedication to preserving the past and giving it a place in the present and future. They continue to manage and work on the farm and now enjoy their grandchildren and teach them about their family and farming heritage. Jase is involved in the farm's operations and, among his varied duties, plans and plants the corn maze each year. Mary Morgan became part of another Williamson County Century Farm family when she married Charles Hatcher of the Hatcher Family Dairy.

The Gentry Farm, also known as Pleasant View, and its owners have been featured in many articles and publications. Allen and Cindy generously share their farm with students, teachers, camping groups, and visitors, as well as other farmers who consider offering similar agritourism activities on their land. They and their family continue the legacy of previous generations by raising crops and livestock, and educating all who visit the farm about the necessity of agriculture.

Wagon rides, learning activities, pumpkins, and so much more are part of a visit to the Gentry Farm (Pleasant View) in the fall.

Reams-Jefferson and Jefferson Farms, 1854

The Douglass-Reams House, built in 1824, is listed in the National Register of Historic Places. The house was sold to Robert Reams in 1854 when he purchased the farm previously owned by the Douglass family. (Photograph courtesy National Register of Historic Places website)

Henpeck Lane is a convenient connecting route between the Lewisburg Pike (U.S. Hwy. 231) and the Columbia Pike (U.S. Hwy. 31). For many years, Douglass Road was the name of this now well-traveled byway south of Franklin. Rev. Thomas Logan Douglass and his wife, Frances, both born in North Carolina, moved to Williamson County around 1824 when her father, John McGee, gave the couple a large tract of land. On it, they built a central hall, one-and-a-half-story brick residence around 1828. Both McGee and Douglass were Methodist ministers. Douglass Chapel was built about 1853 at the junction of Douglass Road and the Lewisburg Turnpike, when several families in the area formed the church and named it after Douglass. In later years, a one-room school was built nearby, which existed until the 1960s. Dr. Samuel Henderson, in his diary that was edited and included in the *Journal* of the Williamson

County Historical Society (#33), records the death of Frances Douglass Love on June 22, 1852. Later that year, Dr. Henderson notes that he bought the Douglass church lot from her estate for $60.00. A fascinating letter included in Henderson's diary is from a former Douglass slave reporting on his journey and his arrival, along with his wife, to take up residence in Liberia in 1854.

After her husband's death, Mrs. Douglass married Joseph Love. Following her death in 1852, her heirs sold the property to Robert Reams in 1854. Robert and Elizabeth North Reams managed about 400 acres and raised tobacco, wheat, and corn, along with cattle and poultry. Skirmishes occurred on the property during the Civil War, and Federal troops set fire to the detached kitchen, but Robert, though ill with typhoid fever, extinguished the flames before they spread to the house. In 1920, a devastating tornado destroyed the kitchen Robert Reams had saved, as well as Douglass Chapel, a substantial brick building. In her book *Historic Williamson County: Old Houses and Sites,* Virginia Bowman includes several stories handed down in the Reams family.

This is believed to be Sallie Reams Jefferson with two daughters and a son, along with his pony, in front of their gable-front and wing home, a style popular in towns and rural areas from the 1870s and into the first years of the twentieth century.

Sallie Reams and her husband, Jonathan Willard Jefferson, Sr., acquired 122 acres of the farm in 1890. It is from this marriage that the Reams-Jefferson line of ownership continues to this day. Sallie and Jonathan are buried in a small family cemetery on the property. One of their sons, Robert Reams Jefferson, married Virginia Carson Jefferson and acquired 50 acres in 1931. On Robert's death, he left shares to heirs, including his wife and Caswell M. Jefferson. Caswell M. Jefferson died in 1977, and his shares and other tracts eventually became the property of his wife, Joyce M. Waggoner Jefferson. They were the parents of Caswell M. Jefferson, Jr. and Susan Jefferson (Hardcastle). Throughout the twentieth century, the family owners maintained working farms with dairy and beef cattle, fruit, grains, and tobacco until, like most farmers, they chose the buyout offered by the United States Department of Agriculture in 2004.

The current owners of the Reams-Jefferson Farm are Caswell M. Jefferson, Jr., and his wife, Kristi Roberta Hill Jefferson. Acquiring 45 acres in 1995, he and Kristi live in a 1906 home, and their sons, Adam and Caswell M. Jefferson III, and their families also live on the family farm.

Susan Jefferson Hardcastle acquired nearly 40 acres of the original farm in 2019, following her mother's death in 2018. She named her property Jefferson Farms, which follows the same history as Robert and Elizabeth Reams. Also making their home on a portion of the ancestral farm are her daughter, Stephanie, her husband, Jeremy Johnson, and their children, Jefferson, Julianna, and Jacqueline May.

The Reams-Jefferson and Jefferson Farms have preserved their history and farm life in a part of the county that is now highly developed with residences, businesses, and new roads. Family members appreciate the long history of their ancestors on this property and plan to maintain their presence on Henpeck Lane.

The grave of Rev. Douglass, who died in 1843, is marked by a tabletop slab, fully inscribed, though weathered and nearly illegible. The grave of his wife, Frances, is beside his monument. Both are in the historic Rest Haven Cemetery in Franklin.

William Steele Farm, 1859

William A. Steele, son of the founding couple, and his wife, Elizabeth, were married in 1850 and completed their house around 1855. Often referred to as an I-House, the two-story frame house with central passage and Greek Revival portico was a popular style during this period. Elizabeth lived in the house until she died in 1911.

The Steele and Bond families in the Bethesda area are connected by generations spanning well over two centuries. A large tract of land originally owned by Moses Steele and John Bond is the origin of the two Century Farms. The property designated as the William Steele Farm was sold by Tom Bond to William Alexander Steele of Virginia. Steele and his wife, Susan Moore Steele, moved to the Bethesda area in the early years of the nineteenth century. Their son, William Alexander Steele, Jr., was born in the county in 1827 and inherited the farm that bears the Steele name.

William A. and Mary Elizabeth Steele

In 1850, William Alexander Steele, Jr., married his cousin, Mary Elizabeth Steele, and, circa 1855, they built the house that is listed in the National Register of Historic Places. The Steeles were slaveholders and managed a diverse operation of row crops, livestock, and foodstuffs for those who lived and worked on the plantation. During the Civil War, William A. Steele was in the Fourth Tennessee Cavalry. Mary Elizabeth relied on slaves and family to keep the farm in production. In *Historic Williamson County*, Virginia, Bowman gives a detailed account of a visit by Federal soldiers to the home in search of Steele, who was hiding but not found. Following the war, the Steeles added the ell to the house. Buildings dating before 1860 include the separate log kitchen, log chicken coop, barn, and the remnants of a log slave dwelling. Over the years, other barns and buildings were added to support the farming operation.

Cora Steele (1866-1915) acquired 169 acres of her parents' farm. Married to William Cicero Bond, the couple and their three children, William Howard, Mary, and Loreen, raised cattle and row crops while living in the two-story farmhouse. William Howard Bond, born in 1899, was the next generation to own his family's property. A veteran of WWI, he wrote letters in 1918 from New York to his father and sisters asking about the family and telling of being sick but getting better. He served overseas from October 1918 until July 1919. After returning, he and his wife Eva managed the farm. Their son William Franklin "Banny" Bond was born in 1927

"Banny" was a Korean War veteran and a familiar face to children and parents on his routes as a Williamson County school bus driver for over 55 years. Bond was also a member of the Farm Bureau's Board of Directors. His wife, Margaret Vantrease Bond, was a teacher's aide with the county school system for twenty-four years. The Bonds were active members of the Bethesda Presbyterian Church. When Banny Bond died, the farm came to his widow and their daughter, Jane Bond Giles. With Mrs. Bond's passing in 2020, Jane became the owner. Here, she and her husband, David Giles, operate and manage the farm where they raised their sons, Corey and Gary.

Jane, David, and Corey manage, operate, and actively work both the William Steele Farm and the Mary and Loreen Bond Farm (1897), which is across the road and was part of the original farm. Knowledgeable and current regarding incentives for farmers, all three have participated in the Tennessee Agricultural Enhancement Program, and the farms are within the Tennessee Green Belt, which preserves agricultural, forest, and open space land. Of this designation, the Giles agree, "It truly helps farmers and allows them to continue to farm." The family raises hay, Black Angus, and vegetables.

When asked what these family farms mean to her, Jane Bond Giles quickly explains, "This is my life and I love it. I don't know any other way of life beyond farming, and I wouldn't sell for any amount of money!" Recognizing that farms have been sold, sometimes needlessly, because of a lack of succession plans by owners and family members, the Gileses have put in place legal documents to ensure the land remains within the next generation.

To sum up her feelings about living and working here all her life, owner and matriarch Jane Bond Giles gives the benefit of her wisdom, "This farm, this land, is not ours. It is God's land, and we are just taking care of it for Him. He has put us here to do His work—raising grains, fruits, and vegetables, and taking care of animals. Each day when we go out to work, I pray for safety and for guidance to do what He wants us to do. I want to take care of His land."

William Cicero Bond (1862-1928) and Cora Steele Bond (1866-1915) and their family, including children William, Mary, and Loreen gathered in front of the main residence completed by Cora's parents, Elizabeth and William A. Steele.

The log kitchen, dating to about 1850

Outbuildings on the William Steele Farm include the 1890 frame smokehouse and the earlier log chicken house

FARMING TRANSITIONS AS MODERN AGRICULTURE EMERGES 1869-1899

Disc Harrow with Iron Frame.

By 1869, the Civil War was officially over, and farms in Williamson County and across the South were rebuilding. Some farm families, however, could not continue farming for various reasons. So many men died in battle or in field hospitals from infections and diseases. Those who survived, often with wounds that prevented farm labor, returned home to find buildings, fences, and pastures destroyed. Depending on their circumstances, some women with young children who had held on to the farm during the war years now realized they could not continue farming, and the land was sold. Hardly any livestock, poultry, grain, and vegetable stores, or orchards had survived the needs and hunger of clashing armies. Without slave labor, some landowners could not or chose not to continue farming. A reordering of the agricultural society that had existed since early settlement now began in earnest.

Former slaves and their children born after 1865 could live and work wherever they could to survive and flourish as citizens of a very different United States. Some enslaved men had escaped early in the war to the Union Army lines and lived in contraband camps, which were in each of the three Grand Divisions. From here, men joined the United States Colored Troops (USCT), which, by 1866, accounted for 40% of the Tennesseans serving in the Union Army. At least sixty men from Williamson County served in the USCT. Three contraband camps were in Nashville, and one was in Brentwood. A few of these camps became the foundation of neighborhoods for freed men and women and the place where their businesses, churches, and schools began to emerge during and after the war.

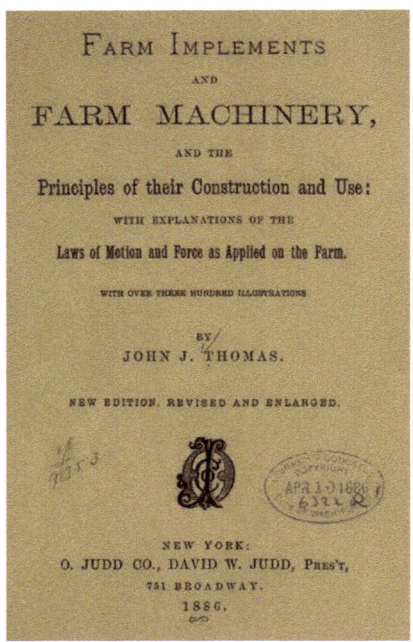

Part of the preface to this detailed and technical 1886 manual informs the farmer that within are both illustrations and descriptions of new and improved options in equipment and tools, the principles behind their construction, and how to use and maintain the machines safely. Thomas estimates that the combined cost of farming equipment across the United States in the late 1880s exceeds a thousand million dollars, with an equal amount expended on labor associated with the machinery. (Publication from the Library of Congress)

Many men and women chose to leave the farms where they had been slaves and move to urban areas, a trend that continued throughout the nineteenth century and for subsequent generations through the twentieth century and to the present. For others, the life of servitude was so deeply ingrained, and the concept of being allowed to make decisions was nearly incomprehensible after generations of bondage. For many former slaves, though now legally free, they were unable to make a living because of continued constraints and prejudices that kept them from progressing. Few Black people could afford to purchase their own plots of land, though some records indicate that a small percentage received plots from their former owners. Some individuals and families chose to remain in place on land they knew as tenants and sharecroppers. This arrangement, depending on the landowners and the skills of the individuals, could benefit both parties, but most often, the owners. Tenancy or sharecropping was a step beyond slavery, but in some cases, a very short step. Poor whites, with little hope of land ownership, also entered into tenancy and sharecropping agreements.

Because of livestock losses during the war years, farmers had to purchase stock from other states and countries. For example, Richard Ewell of Spring Hill, just across the Maury County line, imported the first Jersey cattle to the South soon after the war. With them came the Campbell family from Scotland, who established nearby Cleburne Farm, which operates today as the oldest Jersey dairy in the state and one of the oldest in the country. Jerseys, named for the Isle of Jersey in the English Channel, were prized because of their outstanding production of rich milk. This breed, along with Guernsey cattle, also from the Channel Islands, and the Holstein Friesian from the Netherlands, played a significant role in the nationally recognized dairy industry that began to emerge in Williamson and surrounding counties.

In Goodspeed's *History of Tennessee*, he notes the rapid increase in production of row crops and livestock that occurred in counties in the first years following the Civil War. He records that bushels of corn produced in Williamson County increased from 1,010,443 in 1870 to 1,439,445 by 1885. During the same period, bushels of wheat moved from 227,294 to 315,966 while oats jumped from 99,933 to 585,522 bushels. Significant increases in livestock from 1870 to 1885 were: horses and mules – 10,314 to 11,442; cattle – 6988 to 12,906; and swine – 41,703 to 43,132.

Wood's Engine on Wheels, with Pipe Folded Down.

Lengthy chapters discuss types of plows, new and improved, along with illustrations of implements including the disc harrow, smoothing harrow, corn huskers, corn planters, mowers and reapers, hay loaders, and steam engines – all designed to decrease manual labor and increase farm production.

The recovery and increase in farming output, on both large and small holdings, can be partially attributed to the rapid and diverse innovations in farm equipment and tools, as well as their increased availability. The industrial age, most closely aligned in the South with the arrival of Whitney's cotton gin in the early 1800s, was interrupted by the Civil War but gathered momentum in the last quarter of the nineteenth century. Horse-drawn seed drills made for quicker planting of large fields of row crops. Steam-powered tractors and threshers allowed for greater production and easier harvesting, as did mechanical reapers and binders. These machines saved time and labor, accounting for the increase in grain production. Still, horses and mules were integral to agriculture. Steel plows replaced wood and iron ones, and cultivators, disc harrows, and wagons all relied on horsepower well into the twentieth century. Catalogs of tools and machinery, as well as farm magazines, increased in number, making their way into the hands of farmers looking to expand, improve, and adjust to new options. With this new machinery came the farmer's responsibility to learn how to take safety measures and care for their investment, whether an improved plow or a mammoth steam-powered engine. In this section, the stories of the Century Farms founded during this period represent the transitions in farms, farming, and farm families, which resulted in agriculture firmly placed as the number one industry in Williamson County and Tennessee.

COUNTY LINE FARM, 1869

CRYSTAL VALLEY FARM, 1869

BOND FARM, 1870

MAPLE CREST FARM, 1870

SMITH BROTHERS FARM, 1878

SULLIVAN FARM, 1881

HUNT – BEASLEY FARM, 1886

JOHNSONGRASS OR LAMPLEY FARM, 1886

CHARLES GENTRY FARM, 1887

LOREEN AND MARY BOND FARM, 1887

CEDAR CREEK FARM, 1888

BARKER'S HILLVIEW FARM, 1891

WILSON FAMILY FARM, 1893

EASTVIEW, 1897

LONG VIEW, 1897

PEWITT FARM, 1897

County Line Farm, 1869

The log house, dating from 1869, remains a constant backdrop for the farm's activities. The current dwelling, built in 1927, will soon reach its century mark.

In 2019, Ben and Mary Linton Little invited family and friends to celebrate the 150th anniversary of County Line Farm. As the name indicates, it is located in both Davidson and Williamson Counties and is a landmark property on Big East Fork Road. The town of Linton is mentioned in the Fernvale entry of *Back Home in Williamson County* (Pewitt, 1986) and is still noted on maps. Silas Linton and his wife, Katie Anderson Linton, both born in the Linton community, acquired property in 1869 and built a log house in which they and later generations lived. A barn, still in use, is also the work of Silas from that time.

Hooper Linton, married to Ida McPherson Linton, was the second generation to own the farm. He and his brother Will built the "new" farmhouse in 1927 just in front of the log dwelling. The baseboards and door frames were all made from chestnut that was harvested off the farm—something that could not have been done much later because of the blight that destroyed these trees in the mid-to late 1930s. The house is a replica of the house in which Ida was raised on Big East Fork Road. As well, nearly fifty years after his father built the main barn, Hooper added the "mule barn," which is still in use and in good repair. Both barns are featured in *Barns of Tennessee* (2009) and *Barns of Williamson County* (2019) and are the subject of many photographs made by travelers on the rural and scenic road.

Lloyd Linton and his wife, Frances Edmondson Linton, greet family and friends at the 2014 ceremony to name the bridge over Highway 96 West at the South Harpeth River in honor of the lifelong farmer and WWII veteran.

In 2014, another milestone celebration for family and friends took place when the bridge over the South Harpeth River on Highway 96 at Old Harding Pike was named the "Lloyd Linton Bridge". At the time, Lloyd, the son of Hooper Linton and Ida McPherson Linton, was 94 and had lived his life on the family farm except for the years when he was in the United States Army during and after World War II. The proclamation from Metropolitan Davidson County noted that "in recognition of Lloyd Linton's lifelong commitment to preserving the farming way of life in Davidson County, it is fitting and proper that the bridge on Highway 96 over the South Harpeth River near Old Harding Pike be named in his honor." The ordinance also recognized Frances Edmondson Linton, Lloyd's wife of nearly 67 years, for her joint commitment and management of the farm. Frances, a graduate of David Lipscomb College, was a schoolteacher for twenty years as well as a homemaker and farmer. The Lintons were active members of South Harpeth Church of Christ, and Lloyd was an elder to that congregation for 30 years.

Mary Linton Little and her husband, Ben, gathered with their children and grandchildren in front of the first tractor purchased for the farm. The 1869 log house is included in this photograph of the fourth, fifth, and sixth generations. From left are Andrew, Elizabeth, Mary, Ben, Britt, Shelley, Tess, Hudson, Ben, and their farm dog, Wrigley.

Mary Linton Little and son Andrew receive the 150-year sign at the Williamson County Fair Century Farm event.

Following the deaths of Lloyd in July of 2015 and Frances in September of 2015, their daughter Mary Linton Little became the fourth-generation owner of the property along with her husband Ben Little. Mary was raised on the farm, and she and Ben live in the farmhouse built by her grandfather and great-uncle. They are the parents of Andrew and Brittain, who learned about farming and family history from their parents and grandparents. The crops and livestock have varied over time, though mules have generally been in residence at County Line Farm. Today, cattle are raised along with grains, vegetables, and an abundance of flowers in season.

Mary writes from her perspective as the keeper of the history of the family and farm, "As my age increases, this farm has a deeper meaning for me. I often think of the manual labor required to clear the fields and create the pastures that cattle enjoy today. I know that my Daddy loved every inch of this farm and was familiar with every hill and gully. Even though I have not followed a mule through plowed fields and ridden the old International Harvester over every inch of the cultivated fields, my love and admiration for this land increases with each day that passes." She explains that, in time, the farm will pass to their sons, the fifth generation "who want to keep the farm in the family and for that we are forever grateful."

Top: Silas Linton built the wood-frame barn around 1869. Bottom: His son, Hooper Linton, built the "mule barn" about 50 years later.

Crystal Valley Farm, 1869

The current generations of the McCanless family received the 150-year sign for Crystal Valley Farms at the Century Farms dinner at the Williamson County Fair in 2019.

During the first years following the Civil War, a number of farms were established as Williamson County began recovering and land changed ownership. In 1869, James Thomas Carroll McCanless purchased the 423-acre Copeland Farm in the northeastern part of Williamson County between Triune and Nolensville. Growing cotton and small grains along with raising cattle, swine, and sheep, he named his farm and home Crystal Valley Farm.

According to family records, James was a religious man and is credited with founding two churches. He first married Susan Jane Lovell, and after her death, married Elizabeth Luviney Coleman. He fathered 11 children, nine of whom survived him.

The family of Ardeen McCanless Johnson, circa 1940s, at the farm she owned and managed.

Before James died in 1884, his six oldest children were given their share of the land. The remaining acreage was to be inherited by his widow until her death or the end of her widowhood. After her remarriage in 1885, the three youngest children, Ardeen, Nina June, and D. Brown, filed a successful lawsuit against their mother to receive their share.

Ardeen married William Hazlewood Johnson in 1885, and they lived on the land left to her by her father. Her husband died in 1895, and she married his brother, James Knox Polk Johnson. Ardeen operated the farm until she died in May 1946. Her son by her first marriage, John Johnson, took over in 1946.

In 1970, James Caldwell McCanless, Sr., the great-grandson of James Thomas Carroll McCanless, purchased the property from John Johnson and incorporated the operation as Crystal Valley Farms, Inc., serving as its president until he died in 2011. James and his wife, Barbara Jean, were active in the community and lived on the farm while raising their children: James Jr., Robin Carol Thomas, and Jonathan Lee. Barbara passed away in 2022. Jon McCanless currently lives in the homeplace and is the current president of Crystal Valley Farms, Inc. He advises that the decision was made about three years ago to place 100 acres in a permanent easement with the Tennessee Department of Transportation (TDOT) to return the farm's streams to pristine water condition. Over the years, row crops and livestock damaged the streams, but working with TDOT on "stream bank mitigation" will restore those features. While it is a slow process, McCanless reports that wildlife is already returning to the land, and aquatic life is returning to the streams. Additional acreage is being given a much-needed rest from production while Todd Thomas, Robin's son, farms other McCanless family parcels.

Barbara Jean Little and James McCanless lived, worked, and raised their family on his ancestral farm.

Bond Farm, 1870

The agricultural landscape of Bethesda, in the southeastern part of the county, is both scenic and excellent for raising livestock and hay. Several Century Farms are clustered in this area.

First associated with the Scales, then Grigsby families of Bethesda, the Bond name came with the 1936 marriage of Leo Ratcliff Grigsby to James W. Bond, Jr., whose family was among the earliest settlers in that area. Leo's grandparents were P.D. and Mary Scales, who established a farm of about 80 acres in 1870. The founding couple raised beef and dairy cattle, swine, poultry, and vegetables on 80 acres. Adding 20 acres in 1897 for $200.00, the deed describes the land as bordered by Rutherford Creek that "meanders to the old meeting house." P. D. Scales operated a general store at the intersection of Bethesda-Cross Keys Road and Bethesda-Duplex Road for fifty years.

Ella Scales, daughter of P. D. and Mary, became the second owner of the farm in 1918, purchasing her siblings' shares of the land. She and her husband, Charles F. Grigsby, had a large family, including Leo, Ethel, Scales, Charles, Marion, Ella Frances, Catherine, and Harry. They raised many of the same crops and livestock and added tobacco.

When Ella died in 1957, Leo acquired 50 acres of the family farm. She married James William Bond, Jr., and remembered that they did not have a honeymoon because "he had a dairy, and he couldn't get off much." Leo's full and fascinating life and memories are chronicled in *Bethesda and Surrounding Communities,* authored by Rick Warwick (Williamson County Historical Society, 2023).

James William Bond, Jr., and Leo Grigsby Bond owned the farm that was founded by Leo's grandparents, Pleasant and Mary Scales, in 1870.

Leo attended school at Bethesda for 12 years, then went to the Tennessee State Normal School in Murfreesboro (now Middle Tennessee State University) and furthered her education at Peabody and Trevecca, and through extension courses at the University of Tennessee. She began teaching at nineteen and rode horseback daily to Simmons Hill, five miles away; she taught in county schools for 45 years altogether. Leo and James were the parents of James, Charles, and Dan, who grew up working on the farm, with livestock, including dairy cattle. After James died in 1967, Leo became the sole owner of her ancestral farm. She sold 19 acres to her son, Charles, and his wife, Carol Allen Bond, in 1974. After Leo passed away in 1997, Charles and Carol received the remaining 31 acres of land from her estate. With their sons, Robert and Allen, and Allen's son, Connor, they farm the Bond and Blue Grass Century Farms.

1907--1997. Generations of local students learned from longtime teacher Leo Grigsby Bond, whose life spanned most of the twentieth century.

Maple Crest Farm, 1870

Walter and Kathleen Ogilvie enjoyed riding over their farm while checking livestock.

Located on the Arno-Allisona Road in College Grove, Maple Crest Farm is known for its historically strong ties to the breeding, showing, and sales of the Tennessee Walking Horse. On a farm that dates to 1870, Walter William Ogilvie became the owner of over 500 acres in 1920. Walter was an early and leading breeder of Tennessee Walking Horses and a founder of the Tennessee Walking Horse Association in 1934. William Harris and Annie Lou Ogilvie, Walter's parents, were the farm's founders. He was raised on the farm and learned early in his life how to breed and care for fine livestock. Walter attended Branham and Hughes Military Academy in Spring Hill and the University of Tennessee. He was an elder in the College Grove Presbyterian Church and a Mason. Married to Kathleen Smith, the couple raised their three children on the farm, which, in addition to being known for its horses, was a diverse farm producing burley tobacco, grains, cattle, sheep, and swine. When Walter died in 1977, the property came to Kathleen and their children.

The family of Bill and Jackie Ogilvie included their three daughters, sons-in-law, and grandchildren.

William "Bill" Harris Ogilvie, son of Walter and Kathleen, and his wife, Jacqueline "Jackie" Reid Hawkins, were the next generation to work and live on the farm, along with their three daughters, Mary Helen, Annie Lou, and Lynn. In addition to working on the farm and relishing her role as mother and grandmother, Jackie was a graduate of Vanderbilt School of Nursing and worked as a registered nurse at Vanderbilt Hospital until her retirement in 1962. She was a leader in the Girl Scouts and 4-H Clubs, and, with her family, was active in the College Grove Methodist Church. The farm was certified as Maple Crest Stock Farm during the ownership of Bill and Jackie and was included in *Tennessee Century Farms: An Agricultural Perspective* (West, 1986).

Lynn Ogilvie Woodside is the current owner of the farm and, following her retirement, returned to the farm where she and her husband Mike built a new residence. Now in the family for over 150 years, it is known today as Maple Crest Farm.

Current owner, Lynn Ogilvie Woodside, and her husband, Mike, gather with their family at the farm that has been in her family for more than 150 years.

MAPLE CREST FARM, 1870

Walter W. Ogilvie, an early breeder of Tennessee Walking horses and a founder of the Tennessee Walking Horse Association, shows one of his champions.

A barn with solar panels is a recent addition to Maple Crest.

Whether pasture or row crops, the fields at Maple Crest are verdant.

William Harris Ogilvie and Annie Lou Ogilvie founded a farm that became known for its livestock. Their three children, Walter, Frankie, and Mary were born on the farm. This photograph dates from ca. 1896.

Smith Brothers Farm, 1878

Jeff Holt is baling hay in the far background of the farm which has been in his family for nearly 150 years.

The Smith Brothers Farm takes its name from two sets of brothers, the sons and grandsons of founders Frank Erwin Smith and his wife Sallie P. Cathey Smith. Frank and Sallie began farming their acreage, which is just inside the Williamson County line shared with Marshall County, in 1878. Their sons were John D. and Thomas P. Smith; the next generation of farming brothers were William Franklin and Fred Riley Smith. The founding couple's graves are the first marked in the nearby Smith Family Cemetery, which is still in use.

Thomas P. Smith acquired 200 acres from his father in 1924. He and his wife, Annie Hazelwood Smith, raised twelve children in a three-room house. He built two barns, two sheds, and a smokehouse for the farm's diverse operation. Thomas and Annie are buried in the Smith Family Cemetery. The next generation, brothers Fred Riley and William Franklin Smith, were both lifelong farmers along the acreage on Ash Hill Road.

Jeff and Jennifer Holt with children Clay and Stacey are an active farm family.

Over the years, the farm's owners continued to be active in the Duplex Home Demonstration Club, 4-H, and FFA, as well as the Williamson County Farm Bureau. In 1981, the heirs decided to divide the property legally, but to continue to work the property as one farm. In 1998, Jeffrey Holt, the 2nd-great-grandson of the founders, and his wife, Jennifer, built their home on the farm. In 2008, Holt inherited the acreage east of Ash Hill Road and now farms about 42 acres of the original farm. The family raises beef cattle, hay, and goats. Jeff was in 4-H at Bethesda Elementary and FFA at Page High School and serves on the Board of Directors of the Williamson County Cattlemen's Association. Jennifer was a 4-H member in Arkansas, and their children, Stacey and Clay, were also involved in 4-H. They continue the traditions of family farming while engaging in progressive and modern agricultural methods.

The Smith Brothers, Fred Riley and William Franklin, farmed together for over 50 years.

Thomas P. Smith and Annie Hazelwood Smith had twelve children. He built several of the outbuildings for the farm.

Sullivan Farm, 1881

Sarah Jane Tidwell, the first owner of the farm, and her husband, Andrew Jackson Sullivan, were among several Sullivan families who migrated to and remain in Fairview and other communities in the western part of the county today.

People with the surname *Sullivan* are numerous in the First District of Williamson County. Three Century Farms have Sullivan within their name, and more with the surname are aligned with other farm histories. The Sullivan Farm is the earliest certified Century Farm in this part of the county. In an area called the "Badlands" on Turnbull Creek, two miles from the Hickman County line, is the farm registered in March of 1881 by Sarah Jane Tidwell Sullivan. Married to Andrew Jackson Sullivan, they were the parents of eight children. The family tended an orchard and grew corn and tobacco as well as mules, cattle, swine, and poultry. On the Western Highland Rim, timber was an important commodity, and the Sullivans also had native grasses, including Blue Stem, Little Blue Stem, and Switch Grass.

The second owners were nephews of the founding couple, Allen Judson Sullivan and John Ensley "J.E." Sullivan, who inherited the property in 1915. Two years later, J.E. and his wife, Ida Dora Pewitt Sullivan, became the sole owners. The parents of nine children, seven of whom survived to adulthood, J. E. and Ida managed a diverse farm and operated a sorghum mill. When State Route 100 was built in 1929, the Sullivans and their neighbors helped with the project using mules to pull ground slides. In the 1940s, the Sullivans used terracing, in cooperation with the Agricultural Stabilization Service, to control erosion. A blacksmith shop operated in the 1950s and 1960s, and some tools and equipment from that era remain.

SULLIVAN FARM, 1881

Food is always a serious part of any family gathering. The reunion hosted by John Ensley Sullivan and Nelle Edna Walker Sullivan in the late 1950s was a memorable event.

With their fine flock of chickens are William Houston Sullivan and Lenar Florence Deal Sullivan, circa 1960.

Family ownership changed hands several times among the heirs of J. E. and Ida in the years following their deaths. Eventually, William "Houston" Sullivan, son of J.E. and Ida, along with his wife, Lenar Florence Deal Sullivan, and their daughter, Martha Sullivan, became the farm's longtime owners. In 2005, Houston and Lenar's son, William Earl Sullivan, acquired the farm. Married to Linda Tucker, the couple raised their three children, Vickie, Jeffrey, and Sandy, on the farm, and all were involved in farm chores. After William Earl's death, Linda Tucker Sullivan and Jeffrey became the owners. Today, farm products include cattle, timber, hay, fruit, and vegetables.

Linda Tucker Sullivan celebrates her birthday with her children, Jeffrey, Vickie, and Sandy.

Hunt-Beasley Farm, 1886

Ella Beasley Hunt with daughter Evaline and husband J. Buchanan "Buck" Hunt in front of their home around 1900. The house includes a circa 1830 log portion, and though changes to the porch and roof, along with improvements, have occurred over the years, the house is well-maintained and preserved.

Few travelers on busy Carter's Creek Pike going to and from Burwood and beyond notice the white farmhouse and outbuildings buffered from the road by pasture, but the Hunt-Beasley Farm is still a working farm after nearly 140 years. Many Century Farms remain in the family through ownership by women for one or more generations, and the Hunt-Beasley property is a good example. J. Buchanan Hunt and Ella Beasley Hunt, with their daughter, Evaline, born in 1882, established their farm south of Franklin in 1886. At the time, a log house built about 1830 formed the nucleus of their home, with the family adding rooms and improvements over the years.

Clapboard was placed over the exterior before 1900, and an off-center porch added interest and a place to sit, work, and watch. A treasured family photograph shows the founding couple and a young Evaline in front of the house.

Willie Beasley and her mother, Evaline Hunt Beasley, were the next women to own and farm the house and property. Except for the porch, the home's exterior has hardly changed.

Buchanan Hunt died in 1904, and Ella married Frank Hull in 1908. At her mother's death in 1928, Evaline inherited the farm. She and her husband, Newton Cannon Beasley, continued to operate the farm much as her parents had by growing row crops, grains, and livestock. They also lived in the farm's main dwelling. Their daughter, Willie, became the third-generation owner. Her husband was Joseph Ridley Jones, whose family settled in Hillsboro (Leiper's Fork) in the early 1800s. They were the parents of Alice and Joseph "Jay" William Jones.

A multi-purpose hay and livestock barn was added around the 1950s and in the 1980s another barn was built. A smokehouse was in use from at least the latter part of the nineteenth century and well into the twentieth century. Fields and pastures supported grains, hay, and cattle, and the family always grew vegetables and fruits. In 1985, when her mother died, Alice Jones Sparkman became the third woman in her family to own the property. She married Ollie Jo Sparkman of Maury County in 1954. He worked for John Deere dealers and raised cattle on the farm. Their son is Joseph Hunt Sparkman. The Sparkmans lived and worked on the farm until Ollie passed away at the age of 96 in June of 2025.

Except for the porch, the exterior of the historic home has hardly changed.

Alice Jones Sparkman with her husband, Ollie Sparkman, and their son Joe.

The multi-use hay and livestock barn was built in the 1950s. A newer barn was constructed in the 1980s.

The smokehouse saw decades of service.

Johnsongrass or Lampley Farm, 1886

Molly Lampley, with her husband, Uzebe, founded the farm. She is surrounded by four generations of her descendants.

The First District encompasses Fairview and numerous smaller communities that longtime families continue to identify with today. The district forms a triangle with the base bordering Hickman and Dickson Counties, and the north angle is bound by Davidson County. Residents in that area find it easier to travel to Dickson, especially with I-840, than to come into Franklin. When it was settled in the early 1800s, this area was a wilderness, and it remained so for decades. Until U.S. Highway 100 (the Nashville–Memphis Highway) was constructed, travel was difficult. This is one reason that Fairview was the second town incorporated in the county after Franklin, with its own city government. It was too inconvenient for residents to do business in Franklin. It also had the second high school in the county for the same reasons. Even as late as the 1970s, it was long-distance to call to or from Fairview and Franklin. The construction of Highway 96 West from Franklin to U.S. Highway 100 made an extraordinary difference, quickly bringing residential and commercial development, along with increased traffic along that route. More recently, I-840 added an option for travelers and residents in and out of Fairview and adjacent communities. Today, the First District is one of the fastest-growing sections of Williamson County, and commercial and residential development has changed what was a predominantly rural landscape in the past 30 years. For more reading on the history of this area, *Out There in the First District*, by Rick Warwick (2001), and *Back Home in Williamson County* by Lyn Sullivan Pewitt (1986) each offer excellent insights into the history of the area and its people. Several Century Farms represent the First District.

This house was on the farm when Uzebe purchased it in 1886. He and his family lived in it for several years.

The Lampley name has long been associated with this area, and many descendants of earlier generations still call this place home. The Lampley Farm was founded in January 1886 by Uzebe Lampley, who purchased an existing 162-acre farm on Sugar Camp Road. A house was on the property at the time Uzebe purchased it, and the family made it their home. He and his wife, Molly, had seven children, and the family raised sheep, beef cattle, ducks, chickens, and sold surplus garden vegetables. Their son, Earl Demarquis Lampley, Sr., purchased 131 acres of the farm in 1945 following the deaths of his parents. He and his wife, Ruby, had five children.

Betty Dean Hughes Lampley and Earl Demarquis Lampley

Earl Demarquis Lampley, Jr., born in 1923, acquired three tracts, including the acreage where the homeplace was located, after the deaths of his parents. Known by his middle name, DeMarquis, he was a farmer and businessman who retired after 30 years with Western Electric. Lampley served as a county commissioner from 1960 to 1966 and was a member of the county's election commission from 1976 to 2007. Recognizing that Fairview needed more options for local employment and for the convenience of residents, he worked to bring new businesses to the community. His commitment to improving all aspects of the area continued as he also served as Fairview City Manager from 1966 to 1971. Education and the betterment of local schools in the First District were a part of his successful planning and collaborations during this time.

Demarquis, who died in 2023, was married to Betty Dean Hughes Lampley for 72 years. They were the parents of Vickie, Mark, and Anthony. Their grandson, Stuart Johnson, son of Vickie and Cliff Johnson, acquired 78 acres of the original farm. Stuart raises Black Angus, Charolais, and Herefords along with corn and hay. He is selective in bringing diversity to the farm, maintains a fully operational apiary, and has won several awards for his honey. Johnson was the Mayor of Fairview for 16 years. While always recognizing and appreciating the contributions and longtime ownership of his Lampley ancestors, Stuart chose to change the farm's name to Johnsongrass.

Earl Demarquis Lampley and some of his prized herd.

Charles Gentry Farm, 1887

The Gentry Farm sign symbolizes the proud legacy of this family farm that was established in 1887.

Spanntown Road, north of Triune and just south of Nolensville, does not actually go to a place of that name. The short route meanders for a few miles and then comes to a dead end; the driver then turns left or right to pick up other roads. Charles Gentry, who has either lived or worked on his family's farm for more than 90 years, explained that the road got its name because years ago, members of the Spann family lived in five houses at the end of the road. Mr. Gentry is an excellent source of information and stories about the area and his life on the farm..

Andrew Dellars Gentry and his wife, Mary McCanless, began general farming on 106 acres in 1887. Mary was part of the neighboring McCanless family of Crystal Valley

Century Farm. The Gentrys were parents of ten children, and the family raised both livestock and row crops for their own use and to market.

In 1937, Elliott Brown Gentry, born in 1894, acquired part of the farm. In 1923, he married Minnie Eugenia Green, and they were the parents of Martha Virginia, Jean, Charles, Mary Ruth, Katherine, and Dorothy. During their decades of living and working here, the land supported corn, wheat, hay, tobacco, cotton, cattle, mules, and horses. After her husband's death in 1970, Eugenia became the sole owner, living until age 96.

Charles Gentry, patriarch of the family, visits with his son Wayne, who does most of the farm work now, and his niece Lynne McAlister, daughter of his sister Jean Gentry Mangrum.

Their only son, Charles, worked on the farm most of his life, though he and his wife, Margaret Lampley, whom he married in 1949, lived in Nashville. He was employed by Meadow Gold Dairy for thirty years, though he came to the farm almost every day and on weekends to help his mother manage and work the homeplace. Most evenings, he ate supper with his mother before returning to his home and family in Nashville. After his mother died in 1999, he became the owner of a portion of the acreage. He and Margaret built a house near the farm, where they retired, and he continued to farm.

Their son, Wayne, lives nearby and does most of the work on the farm. He raises cattle and hay, which he sells to other farmers if there is a surplus after feeding his herd. Wayne was in the Air Force and was with the Safety and Security Department at Middle Tennessee State University. His son Charlie also helps on the farm.

At age 97, Charles Gentry observed changes in both Nashville and in the Triune/Nolensville area that he readily admits he could never have imagined. Nolensville is developing very quickly, and the large dairy and livestock farms that made the area such a productive agrarian region are nearly all gone. Spanntown Road is now paved rather than the dirt road of his youth and is rapidly filling with houses and other buildings. The traffic increases each year, and moving farm equipment to take care of the cattle is a daily challenge, explain Wayne and his father. When considering the two generations before him, Mr. Gentry especially appreciates his mother and father and their lifestyle and dedication to family and the farm for most of the twentieth century. When asked what the land means to him, he answers without hesitation, "It is everything to me. I love it!"

The pond and gently rolling hills of the Charles Gentry Farm are part of a rapidly disappearing agricultural landscape in the eastern part of the county.

Loreen and Mary Bond Farm, 1887

Cora Loreen and Mary Elizabeth Bond

The history of this farm parallels that of the William Steele Farm, established in 1859, until 1887. At this time, Cora Steele Bond and William Cicero Bond became the owners of the Steele Farm. Their children, Mary Elizabeth Bond, Cora Loreen Bond, and William Cicero Bond, grew up on the farm and in the main two-story dwelling. When their parents died, William became the owner of the larger parcel and house, and Mary and Loreen inherited 205 acres across the Bethesda-Arno Road. They managed their acreage and raised corn, hay, and cattle. In the 1920s, they built their home which they lived in for many years. Neither sister married.

Their great-niece, Jane Bond Giles, who also owns the William Steele Farm, became the owner of this farm. The house of Mary and Loreen is now the home of Corey Giles, his wife, Jill, and their son, Luke. Also on this farm is a rare four-crib Appalachian log barn that was featured in *Barns of Williamson County*. Jane, her husband, David, and her son, Corey, manage and work both farms.

The house was built for Mary and Loreen Bond.

A rare log four-crib Appalachian-style barn is still used.

Four generations attended the Century Farm dinner at the Williamson County Fair in 2015 when the Mary and Loreen Bond Farm was recognized as a Century Farm. From left are Margaret Vantrease Bond, Jill Giles, Corey Giles, and their son Luke, with Corey's mother and Margaret's daughter, Jane Bond Giles, and her husband David Giles.

Cedar Creek Farm, 1888

The farmhouse, built circa 1895 by W. J. and Catherine Trice, is in the Cross Keys community and is home to the fifth and sixth generations of their descendants.

Brothers Charlie, Sam, and Grover Trice were the children of the founding couple, as was Mary Louise Trice, pictured with her husband, A. F. Hargrove. It is their family who owns and lives on the farm today.

The Trice family, from nearby Bedford County, moved into the Cross Keys community, near Bethesda, when brothers W.J. and J.G. Trice purchased 95 acres at public auction in January of 1888. Their operation included tobacco, wheat, corn, and dairy cows. Always active in the community, the Trice family gave several acres to establish the Cross Keys Cemetery. W. J. Trice built a house for his family that included his wife, Catherine Miranda, and their children, Grover, Charlie, Blythe, and Mary Louise. Mary Louise and her husband, A. F. Hargrove, assumed the farming operation after W. J.'s death, and lived on the home place with their daughter Billie Jean. The family continued to grow grains and operate a dairy. They also built a new barn to replace the original nineteenth-century one.

The children and grandchildren of Diane and Gene Marlin gather at Cedar Creek. From left are: Jesse and Amanda Pilkinton with their children: Will, Caroline, Rob, Hattie, and Nan; Gene and Diane; Trey and Jennifer Marlin Lloyd, and their children, Camden and Ava Claire.

When Mary Louise died in 1987, three generations of women became the owners. Billie Jean Hargrove Lillard, who was married to James Marvin Lillard, is the daughter of Mary Louise; Diane Lillard Marlin, married to Gene Marlin, is the granddaughter of Mary Louise; and Amanda Marlin Pilkinton is the great-granddaughter of Mary Louise. Billie Jean died at the age of 91 in 2022, and Diane and Amanda became the owners. Three generations live on the farm today, and cattle and hay are raised on 89 acres of the original property bought by the Trice brothers. The late nineteenth-century house built by W.J. Trice, now restored, is the home of his and Catherine's 2nd-great-granddaughter, Amanda Marlin Pilkinton, and her family. Amanda's sister, Jennifer Marlin Lloyd, and her family visit often and love the family farm, too. It is Diane Marlin's plan that her daughters will continue the line of ownership by women. She writes, "My hope and desire is to see this little slice of heaven continue in our family and be a place of rest and refuge for my grandchildren in the future, and remember what most of Williamson County was like at one time." Diane and her family are teaching the younger children the importance of their heritage and explain, "We are committed to this farm being owned by the family for generations to come."

This modern house is part of the evolving landscape at Cedar Creek.

Cedar Creek, in the southeastern part of the county, is both a scenic and working farmscape.

Barker's Hillview Farm, 1891

Robert Houston Barker, Sr., and Lottie Lee Lavender Barker were the farm owners by 1926, and their line continues today.

The picturesque community south of Franklin on Carter's Creek Pike was initially named "Williamsburg" for the Williams family. Later, it was referred to as "Shaw" to honor the Shaw family. Today, it is known as Burwood, a name suggested by James Drake Pope, from the novel *Robert Elsmere*, published in 1881 and authored by Mrs. Humphrey Ward. The book became a best-seller in both England and the United States.

In sight of the village of Burwood, which at that time included three stores, along with a blacksmith shop and a sawmill, is the farm on Pope's Chapel Road, founded by Andrew Jackson Crowson and his wife, Telitha Barker Crowson, in 1891. The parents and their children, Lou Ella, Willie, and Prentice, grew corn, tobacco, hay, and cattle on 160 acres.

Not all Century Farms pass directly from parents to children. It was Telitha Barker Crowson's nephew who became the next owner in 1926. Robert Houston, Sr., and Lottie Lee Lavender Barker and their children—Annie Lou Barker (Grigsby), Polly Barker (Duncan), Robbie Barker (Norman), Ruth Barker, R. H. Barker, Jr, and Franklin Neil Barker—lived on 187 acres. The family grew many of the same crops and raised cattle and swine as did other farmers in the area. The family recalls that "R. H. Barker, Sr. was recognized by his neighbors as an outstanding farmer, church man, and civic leader."

Franklin Neil, Sr., and Hilda Stokes Barker became the owners of 135 acres of the original farm in 1967.

Franklin Neil Barker acquired just over 135 acres in 1967 from the other heirs of his father. He was a WWII veteran and a rural mail carrier for 40 years with the Thompson Station Post Office. He and Hilda Stokes Barker were married for sixty-two years, and they were the parents of Franklin Neil, Nancy Barker (Craig), and Sandra Barker (Heyboer). Hay and beef cattle became the primary commodities. Hilda, whom the family remembers as a fabulous quilter, painter, and cook, was a much-loved mother, grandmother, and great-grandmother. After her husband's death in 2006, Hilda continued as the farm's overseer until her passing in 2016 at the age of 92.

The daughters of Franklin and Hilda Barker became the owners of the family farm in 2016. Their brother, Franklin "Frank" Neil Barker, Jr., died in 1991. The sisters manage the farm while Larry Jones works the property, raising hay and beef cattle. Nancy and Sandra recall with fondness growing up hiking, riding horses, and driving their old jeep over the land—avoiding the cattle as they grazed. Their children and grandchildren consider it a "treat to go to Burwood, hop into the back of the truck, and explore the farm. We are always thankful to be still able to love the same land as our ancestors."

Franklin Neil Barker, Sr., and his wife, Hilda Stokes Barker, with their daughters, Nancy Barker Craig and Sandra Barker Heyboer.

Franklin Neil Barker Sr. and Frank Jr. work together on the farm. Photo from the mid-1970s.

Sandra Barker Heyboer and Nancy Barker Craig continue the ownership and enjoy the farm with their families.

Wilson Family Farm, 1893

*The hay and livestock barn, built about 1900, now serves as a focal point for seasonal activities and events. It was featured on the cover of **Barns of Williamson County**, which was published by the Williamson County Historical Society in 2019.*

In recent years, farm families have recognized that providing experiences for visitors and hosting events on their land is a viable way to both educate the public and generate income. The Wilson Family Farm in the Bethesda community is owned by Aaron Wilson, who, with his wife Lynn and their family, share their farm with others on a seasonal basis. "We wanted to create something that would give people a chance to come out and enjoy the country, disconnect for a moment, and get away from the hustle and bustle," explains Aaron.

The Wilsons have been the keepers of family photographs and stories, as well as buildings and farming traditions, for over 130 years. In 1893, John B. Bond purchased 140 acres, the same land that remains in the family. John and his wife Emma Catherine Sprott were the parents of Leonard, Lucille, and Gladys. The children attended

Bethesda School, at the corner of Bethesda and Arno Road, where the Bethesda Market is today. The family remembers that Catherine walked by the "pony cart to see that Leonard got to school every day." Catherine died in 1902, leaving John a single father who did his best. Gladys recalled that she and her sister would run and hide under the bed so their Papa could not reach them to comb their hair. The two photographs, however, show the young ladies beautifully dressed with combed hair and bows. The family raised cattle, swine, and chickens along with row crops, including tobacco. Tenant farmers worked with the Bonds and lived on the farm.

The elegant two-story house was home to the John B. Bond family. With elements of the Italianate style and millwork at each gable, it is a fine example of the changes in housing styles made possible by evolving construction techniques and mass-produced materials in the last quarter of the nineteenth and early twentieth centuries.

Leonard Bond followed his father as owner of the farm. Leonard attended Battle Ground Academy in Franklin, where he was president of his senior class and played football and basketball. World War I intervened, and his life was never the same. The family tells how he was listed as missing in action and presumed dead. This was almost his fate because he was sprayed with mustard gas and left on the battlefield. He was found alive, however, and taken to a nearby hospital by the Red Cross. Eventually, he returned home to the farm and married Elise Core. Their daughter Dorothy named the farm Maple Lawn Farm in the 1940s because of the maple trees lining the driveway.

In 1999, Aaron C. Wilson, the great-grandson of the founders and grandson of Gladys Bond Wilson, acquired the house and farm. He and his wife, Lynn Chester Wilson, and their sons, Riley, Lucas, and Landon, and daughter, Samantha, make their home on the farm. They raise a variety of crops, including corn, soybeans, pumpkins, popcorn, and wheat, as well as chickens. Fully engaged in agritourism, the Wilsons offer a wedding and event venue, seasonal hayrides, a pumpkin patch, a corn maze, Christmas tree sales, flowers, honey, and canned goods, among other items. Aaron recalls how the farm "was a joy for us when we were growing up, and for our children, and it's more important than ever that we offer that joy to others."

WILSON FAMILY FARM, 1893

Leonard, Gladys, and Lucille Bond, children of the founders, are in their trap, pulled by Black Beauty.

Riley, Landon, Samantha, and Lucas, the children of Aaron and Lynn Wilson, are in the barn loft.

The farmhouse is splendid in all seasons.

Eastview, 1897

Though the Pope house dates to the early 1800s, Eastview Century Farm dates to 1897, when Delilah Lavender Shaw acquired 109 acres.

Eastview, from which the farm takes its name, is one of the oldest dwellings in the county and is located in the southernmost part of the county, near the Maury County line. The original part of the house is a two-story log building and, unlike others that were more primitively and quickly constructed, has wainscoting, chair rails, and plaster walls. Built by John Pope, with the help of slaves, it is listed in the National Register of Historic Places. From Granville County, North Carolina, Pope settled here not long after Williamson County was established in 1799. He donated land for and was involved in the building of Pope's Chapel in 1818. At his death in 1829, Pope owned at least 1,000 acres, and his will lists 15 slaves. The farm includes burying grounds for the Popes and those enslaved persons who worked and lived at Eastview. Altogether, this complex represents the history of this community from early settlement to the present day.

The current line of family ownership of the property is traced to Delilah Lavender Shaw, who acquired 109 acres in 1897. She and her husband, William A. Shaw, were from families that were already long-established farmers and active citizens in their community, their church, and local government. During the couple's ownership and that of their son, William E. Shaw, they raised dairy and beef cattle, as well as corn, hay, and tobacco. When W. E. Shaw died intestate, the probate record gave a most useful and detailed description of the property, including the house, outbuildings, landscape, and West End School.

William Augustus and Delilah Lavender Shaw

Annie Lou Barker Grigsby and L.B. Grigsby were the third-generation owners of Eastview.

Robert H. Barker, Sr., and L. B. Grisgby purchased the property in two tracts in 1936 and 1948. Barker's wife, Lottie, was a niece of both Delilah and W. A. Shaw. Grigsby and his wife, Annie Lou Barker, operated a Grade B dairy. Electricity came to the farm in 1940. A year later, he purchased his first and only tractor. The two Grigsby children, Judy and Jack, were very much involved in farm work and were active in 4-H. Judy raised two steers each year to pay for college.

Judith Grigsby and her steers

Judith Grigsby Hayes became the sole owner of the property in 2023 and is the fourth generation to live on the farm. In her efforts to document family and community history, she began improving her property, including restoring the schoolhouse. Because of her lifelong residence in the community and her recognition of the need to preserve history, Judy sponsored the Williamson County Historical Society's publication of *Burwood and Beyond* by Rick Warwick in 2024. Likewise, she initiated and is the sponsor of this publication on the county's Century Farms. Judy manages her farm while continuing her significant contributions to the betterment of Williamson County.

West End School in Burwood

Judy's recent restoration of West End School, which operated as a public school from 1885 to 1915, captured the attention of Burwood residents and others in the area. The preservation of this survivor of the not-too-distant past, though worlds apart from the county's new schools, is due to her efforts and vision. As a former teacher, Judy is an advocate for educating all ages about the history of people, traditions, places, buildings, and the importance of agriculture and farmers.

A former long-time and diligent county commissioner, Judy remains an active member of the county fair board, among other responsibilities and volunteer positions.

She was married to the late Jim Hayes, radio personality and executive, for 55 years. Together, their resolute and significant efforts on behalf of Williamson County continue to benefit the entire region in countless ways.

Long View, 1897

Long View, Mitchum House

Long View is located just off U.S. Highway 31 in Spring Hill, in the southernmost part of Williamson County near the Maury County Line. It is associated with the Pointer family, who came from Virginia to settle just north of Spring Hill in the 1820s. With them was their young son, Henry, born in Virginia in 1822. He was first married to Martha Caldwell Pointer, daughter of Dr. St. Clair Caldwell, one of the first physicians in Spring Hill. She died in 1852. Pointer enlisted in the Confederate Army and was commissioned as a Captain. He was a prisoner of war for several months before being exchanged and sustained wounds in different battles and skirmishes, but soldiered to the end of the war. Pointer returned to the farm and married Virginia "Jennie" Brown, also of Spring Hill, and daughter of Col. Hugh Brown. They had one son, Henry Strange Pointer. After Capt. Pointer's death, two hundred additional acres across from their home on Columbia Highway were acquired by Jennie in 1897. Jennie owned and managed the farm with Henry and his wife, Mattie "Patsy" Campbell Pointer, daughter of Captain McCoy Clemson Campbell. Family tradition recalls that Jennie became ill in March 1929 and died at age 82 after spending a day in the rain with a shotgun to prevent highway workers from cutting trees to widen the road in front of her house. Later, the right-of-way was taken from the Pointer property on the other side of the road. Her only child, Henry, died seven months later.

Though Patsy and Henry had no children, when Patsy's older sister, Mary Louise, who was married to Horace Polk, died in 1901, leaving four children under the age of six, the couple helped raise the children. The boys, Allen and Horace, lived with their father nearby on Beechcroft Road, while the girls, Mary and Alice, moved to their grandparents' farm, Cleburne Jersey Farm, on Sugar Ridge. Mary returned home from Birmingham to help her aunt and sister care for her uncle Henry. After his death, and her marriage the following year to Millard Mitchum, Sr, she and her husband moved into Long View, making their home with Patsy. They were the parents of Alice Ann and Millard 'Bud' Mitchum, Jr.

Henry and Patsy Pointer helped raise the children of her sister, Mary Louise Polk, who died in 1901.

Bud Mitchum

Upon Patsy's death in 1965, the farm became Mary's, and upon her death in 1983, Long View was divided between Alice Ann and Bud, who retained ownership of the 1904 home within his parcel. Bud graduated from Spring Hill High School and attended the University of Tennessee. He served in the U.S. Army in Colorado and Germany and returned to the farm, where he remained until he died in 2012. Of the original 1897 farmstead, the family still owns 65 acres in two separate tracts. In his Century Farm application, Bud noted that barns, a buggy shed, and the house remained from the time Jennie acquired the property.

Bud Mitchum was a popular, progressive, and successful farmer who, at one time, owned and operated farms in four counties. He was active in the Tennessee Farm Bureau, Maury County Co-Op, Tennessee Cattlemen's Association, as well as other agricultural, civic, and community associations. He especially enjoyed sharing his farming knowledge and experience with young farmers. He was a lifelong member of historic Grace Episcopal Church in Spring Hill. When Bud died, his niece and nephew, Alicia and Stephen Fitts, children of Bud's sister Alice, became the next generation to own the farm. Alice Ann's parcel of 41 acres remains within a family partnership controlled by Alicia and Stephen. The homeplace overlooks Highway 31. The surrounding 21 acres are cultivated for hay, while the 42-acre parcel, no longer contiguous, is cultivated for corn, soybeans, and winter wheat.

Alice Ann and Bud with their father, Millard Mitchum, Sr.

Storing hay

Long View Cattle

Alicia, Robert, and Stephen with their grandparents, Mary and Millard

Pewitt Farm, 1897

Built in 1897 by the founding couple, this farmhouse has been home to generations of the Pewitt family, up to the present.

On Green Chapel Road in the western part of the county, Ephraim "E.W." Pewitt purchased 105 acres in 1897 and built a house for his family. He and his wife, Mary, raised four children — Bennie, Earl, John, and Gentry. Like most farming families, the Pewitts were largely self-sustaining, growing a large vegetable garden and keeping cows, chickens, hogs, and mules for their own use and to sell.

Bennie Pewitt acquired the property in 1951. By that time, the farm had expanded to 150 acres, and the products continued to be livestock, hay, and a variety of vegetables and fruits. Bennie and his wife, Velma Harris Pewitt, were the parents of daughters Zelda and Kay. Velma retired as the cafeteria manager for Fairview Elementary School. The Pinewood Ramblers were a well-known local bluegrass band, and both Bennie and Velma played and sang with the group for many years.

Zelda Carson, Kay Vaughn, and their parents, Bennie and Velma Harris Pewitt

Zelda Pewitt Carson became the third-generation owner of the farm in 1998. At this time, about 70 acres were yielding hay. Byron Carvell Carson and Rondall Stacey Carson are the children of Zelda. In 2020, Byron acquired his great-grandparents' property. He manages the farm and grows hay on 48 acres. Byron and his wife, Lisa, live in the county while Rondall Carson, his brother, resides on the farm in the 1897 house built by his great-grandfather.

The multi-purpose barn on the Pewitt Farm is still in use.

The founders and members of succeeding generations of the Pewitt family are buried in Green Chapel Cemetery.

OPTIMISM YOKED WITH REALITY, 1900-1922

In my judgment, complacency with respect to the security and sufficiency of our land is what we have most to fear. — H.H. Bennett, 1928, a pioneer in soil conservation

A lone farmer plowing behind a team of mules is a timeless scene from early settlement into at least the 1970s. Farm Security Administration Digital Collection, Library of Congress

Each generation copes with myriad changes, challenges, and choices, and there were many in the early twentieth century. Innovations and advances in all aspects of society brought a sense of optimism, but it was also a time of harsh realities. The United States endured the effects of the tumultuous and costly World War I

(1914-1918); the Spanish Flu pandemic (1918-1919); and economic depressions in 1920-21, 1923, and 1927, culminating in the Great Depression that began in 1929 and lasted into the 1930s, or, as has been debated, until World War II. The increasing rise in industrialization and urbanization, and the social changes each entailed, brought new and constant pressure on farmers to sell their land for housing, industry, and commercial ventures. Many farms that remained intact through the nineteenth century and even through the economic seesaw of the early years of the 1900s were sold as family members pursued other professional paths, or hardships left farmers with no choice but to sell their land. Other individuals and families, however, were able to take advantage of opportunities to expand their holdings and increase their marketable options.

> In the first quarter of the twentieth century, when the Century Farms in this section were founded, optimism was indeed yoked with reality, but farming could be — and was — profitable. In the new century, agriculture remained the number one industry in Tennessee. Increasingly, though, men and women in farming worked away from the farm for additional income, and farming acreage became smaller. The stories of the farms established during this period represent resilience, adaptability, sound business practices, openness to innovation and new ways of doing things, while also utilizing tried-and-true methods. As with farms established in previous decades, the owners exhibit a love of the land and a determination to farm in Williamson County.

For African American farmers, this period continued to be a challenge in many ways. Still, some former slaves, or their children or grandchildren, finally realized the dream of purchasing and working their own land. At the turn of the 20th century, former slaves and their heirs owned 15 million acres, mainly in the South. By 1920, African American farms numbered 925,000, which was 14 percent of all farms. This was the height of Black ownership of farmland in the United States, which steadily decreased through the following decades for many reasons. In 2007, the USDA reported 47 African American farms of the total of 2,193 farms in Williamson County. This was approximately 2%, which was in line with the national average. Ten years later, the number of African American-owned farms in the county had decreased to 25.

The boll weevil made its appearance in the last years of the nineteenth century and, by 1920, had decimated cotton crops across the South. What had once been the major cash crop for farms, large and small, was destroyed. In light of this reality, the farming industry recognized it must diversify its crops rather than rely on a single crop for annual income. Other plants and breeds of livestock were introduced, along with innovations in equipment, tools, and marketing farm products.

The Progressive Farm Movement of the early twentieth century focused on providing solutions and assistance to farm families, helping them navigate the changes in agriculture and their own lives. It was also an intentional effort to increase productivity through sustainable agricultural practices and education. New methods of farming based on scientific studies and experimentation, innovative and available machinery and equipment, governmental resources, and changes in demographics combined to affect farming and the entire scope of agrarian lifestyle and rural society.

The Smith-Lever Act of 1914 established agricultural extension services connected to land-grant universities, including the University of Tennessee and Tennessee State University (formerly the Tennessee Agricultural and Industrial College). Through a growing network of agents and chapters, information on advances in agricultural sciences, home economics, health, and innovative ideas was taken directly to farming men and women. Home Demonstration Clubs for women (now FCE – Family and Community Education) and 4-H Clubs for children in elementary and high school were established by the Smith-Lever Act. The impact on farming and farming families in Williamson County, Tennessee, and across the country from that time, and continuing today, cannot be overstated.

The dairy industry, which began to emerge in the last half of the nineteenth century, became a lucrative, albeit time-consuming, investment, and many farmers turned to dairying. To accommodate the dairy cows, milking parlors and multi-use stock barns were built. Creameries were established to purchase and pick up milk daily for processing into dairy products. This arrangement provided a steady stream of income for farmers and a constant supply of milk for commercial creameries and dairies as urban areas expanded and grocery stores stocked a variety of cheeses, milk, cream, and ice cream.

Williamson County remained a leader in agriculture. For example, the 1920 census lists the county as second in wheat production among Middle Tennessee counties, with 277,007 bushels from 24,802 acres. Other standard crops included corn and tobacco. The rich soil in most of the county produced an abundance of melons, grapes, orchard fruits, berries, root vegetables, cucumbers, cabbage, and beans. Beef cattle, horses, mules, and swine were raised in increasing numbers along with goats and sheep.

The transition in farming is captured at a wheat threshing at Peaceful Valley Farm in Flat Creek. Workers are using an early tractor, a steam engine, horse-drawn, and mule-drawn wagons. One of the first automobiles in the county is also part of the scene.

BUD'S LONG VIEW, 1900

WALKER FARM, 1900

WILLOW RUN, 1901

HATCHER FAMILY FARM, 1903

SULLIVAN-GIVENS FARM, 1904

PEACEFUL VALLEY, 1905

LUSTER FARM, 1906

BLEDSOE-SULLIVAN, 1906

NICHOLS FARM, 1910

PENN HOLLOW FARM, 1913

WILLOW CREEK, 1915

THOMAS JEFFERSON BYRD FARM, 1920

CRESTVIEW, 1922

Bud's Longview Farm, 1900

The house built in 1878 by J. W. Anderson and lived in by Bettie Bond and Walter Anderson and their family, including son Bud, from 1900 to 1993, was moved to the Grassland community.

From a hilltop overlooking I-840 and Lewisburg Pike, the WSM Tower in Brentwood is visible, as are miles of surrounding landscape. Bud Anderson, in whose memory this Century Farm is named, lived and worked on the farm his entire life. His father, Walter Anderson, bought sixty acres at an auction at the Williamson County Courthouse in 1900. The land was important to Walter because it was part of the estate of his father, Thomas P. Anderson. His brother, J. W. Anderson, built a two-story house in 1878, which became the home of Walter and his wife, Elizabeth "Bettie" J. Bond, and their family in the first year of the twentieth century.

With the six children who survived to adulthood—Bess, Walter, Jr., "Bud," James, Mary Frances, Brownie, and Tommie—the Andersons grew hay, tobacco, grains, and alfalfa while raising hogs and Jersey dairy cows. Walter also had a fondness for horses and was always known to have a saddle mare ready to ride around the farm.

An important year for the Andersons was 1940, when Walter, Sr., was elected to and installed in the Tennessee House of Representatives. In the same year, the farmhouse was wired with electricity as this monumental change became available across the county. During his productive life, Walter also served on the county school board, was a road commissioner, and an elder in the New Hope Presbyterian Church, where he taught a Bible class.

Bettie Bond and Walter Anderson

In 1958, Walter died and left the farm to his children, Bess and Bud. Both were unmarried and had helped their parents run the farm since 1940. They continued to grow tobacco and corn while raising swine, dairy and beef cows, and walking horses. When Bess died in 1965, her brother inherited her portion of the farm. Upon Bud's death in 1992, his surviving siblings and their families inherited the farm and put it up for auction. Recalling how Walter acquired the farm in 1900, Dan Bond, the great-nephew of Bettie Bond Anderson, bid at auction and purchased about sixteen acres of the farm that had belonged to his relatives.

Dan and his wife, Phyllis, a realtor, now own 41 acres of the original Anderson farm. They were both active in 4-H as were their children, Danny and Megan. Dan, who was raised and farmed with his family on the Bond Century Farm in Bethesda, is a longtime employee of the Tennessee Department of Agriculture. He is a volunteer with the Williamson County Fair, the Tennessee Agricultural Museum, Williamson County Farm Service Committee, Williamson County DHIA Board, and the Williamson County Livestock Association. Surrounded by congested roads and development, this property remains a working farm and pastoral landmark.

BUD'S LONGVIEW FARM, 1900

Black Angus cows and calves calmly roam and graze on the hillside of the farm that overlooks the traffic of I-840 and Lewisburg Pike.

Top Right: Needs change over the years, and buildings are replaced to accommodate crops and livestock. With tobacco no longer grown on the farm, a new barn was built for the cattle operation. Left and Bottom Right: On a cloudless day, the WSM Tower, fifteen miles away, can be seen from atop the hill at Bud's Longview.

Walker Farm, 1900

The Walker family — Virgil, Dewar, W.G., Geneva, Era, Benton, and Carl

The restored Triangle School and Jingo Post Office are located in Fairview's History Village. Fairview was known as Jingo from 1889 to 1937, and people born during those years have Jingo listed as their birthplace.

Carl H. Walker, Sr. was born in Jingo in 1916. He attended the local elementary school and earned his high school diploma from Watkins Institute in Nashville. Walker attended what is now Middle Tennessee State University, Vanderbilt, and George Peabody College for Teachers, earning both the B.S. and M.A. degrees from Peabody. In between earning the two degrees, he served in the Coast Artillery. Married to Nancy Peck, the granddaughter of long-time school superintendent Fred J. Page, for whom

WALKER FARM, 1900

county schools are named, the Walkers had six children. Walker taught and coached in Williamson and Davidson Counties for 35 years. Of Carl Walker, Sr., neighbor Stacey Givens, of the Sullivan-Givens Century Farm, said he was "an absolutely kind, infinitely smart, and amazing man."

The restored Triangle School in Fairview's History Village is closely aligned with the Walker family. (Photo by Brent Moore)

Carl's grandparents, William Thomas Walker and Harriet Beech Walker, bought 180 acres from J.T. Sullivan in 1900. The Walkers operated a grist mill and a sawmill for the surrounding families and delivered lumber to the Nashville area by mule-drawn wagon. Their son, William George "W.G.", and his wife, Era Frances Overby Walker, were the next owners. Era taught at Triangle School and was active in the Union Valley Home Demonstration Club. Her husband was a member and director of the Farm Bureau, and their property served as a demonstration site. W. G. also served on the highway commission and helped bring new and better roads into what was a somewhat isolated part of the county. He was a leader in securing Triangle and New Hope Schools for the area. Children could attend local schools through junior high, at which time they were bused several miles to Hillsboro High School (Leiper's Fork). The *Review Appeal* of April 15, 1943, featured an article on the Walker family by Jane Owen, whose valuable "Who's Who in Williamson County" column documented many longtime residents. That article and photographs, along with other families from the western part of the county, are included in the Williamson County Historical Society's publication, *Out There in the First District* (2001), by Rick Warwick.

Carl "Bubba" Walker III, Frank D. Walker, and Anthony Benton Walker, grandsons of Carl Sr. and Nancy Walker, have owned the farm since 2009. On 112 acres, primarily managed by "Bubba", hay, cattle, and timber are the farm's products. Progressive and successful farmers for 125 years, the Walker family continues the strong farming traditions that mark this westernmost part of the county.

Willow Run Farm, 1901

The house at Willow Run evolved over the years to comfortably accommodate each generation.

In 1901, Thomas Harvey (T.H.) Page and his son-in-law, Archer Lee (A.L.) Jordan purchased acreage near the Trinity community, also known historically as Rock Hill. The men, along with their wives, Nannie McClaran Page and Annie Page Jordan, daughter of T.H. and Nannie, raised row crops, Jersey cattle, horses, mules, and vegetables. T.H. Page was a founding member of Trinity Methodist Church. A.L. Jordan was a breeder of registered Jersey cattle. His children and grandchildren participated in showing these animals. A.L. and Annie were the parents of one daughter, Nannie Sue, and six sons, Thomas Lee, Garner, Aubrey, William, Bruce, and Walter.

Walter C. Jordan, the grandson of T.H. and Nannie Page and the son of A.L. and Annie Jordan, acquired the farm in 1945. He and his wife, Emma Ida Wilson, and their son, Walter Lee Jordan, farmed approximately 125 acres on which they grew corn, tobacco, Angus and Jersey cattle, horses, and mules. Walter C. Jordan was an accomplished vocalist, and Emma Ida was an excellent pianist. They played and sang at Trinity Methodist Church for many years, as well as for funerals, weddings, and community events. Walter C. Jordan was also a long-time member of the Williamson County Court, and Emma taught school in Williamson County.

Walter and Sarah Jordan were recognized at the Williamson County Fair in 2012 when Willow Run was certified as a Tennessee Century Farm.

In 1992, Walter Lee Jordan, a World War II veteran, inherited the farm from his father. The generations of the Jordan family have been active in the Home Demonstration Clubs, Farm Bureau, and 4-H. Walter Lee Jordan taught in county schools before serving as Senior Archivist, Director of Records Management, and Director of the Archives at the Tennessee State Library and Archives. Sarah retired from Williamson County Schools and volunteered at the Williamson Medical Center. Married for over 70 years, Sarah died in 2015, and Walter passed on in 2018. The current owner and manager is their son, Dr. Lee Jordan, who leases the land to a neighboring farmer to keep it in agricultural production.

The main barn at Willow Run predates the 1901 ownership of the Jordan family and continues to be used in the farming operation.

Hatcher Family Farm, 1903

Jasper Hatcher, the third-generation owner, displays his service award for "Dedicated Service and Commitment to the Farmers of Williamson County" from the Farm Service Agency.

One of only two African American Century Farms in the county, this farm was established when Meredith Hatcher purchased a farm of 96 acres on Owen Hill Road between Allisona and Arno in 1903. The Century Farm application recognizes the significance of his achievement, stating that "as an African American farmer, in the early twentieth century, his acquisition and cultivation of land stand as a powerful testament to the perseverance and vision in a time marked by social, economic, and agricultural transformation." By 1920, Hatcher had expanded the farm to 317 acres, a challenging time for most farmers, and even more so for African Americans, who continued to face limited options and prejudice during post-Reconstruction.

Hatcher was first married to Rosa Moss Hatcher until her death and then to Vinnie Starnes Hatcher; each was born a slave. The family is traced to Ned Hatcher, born in

1810, and Mariah Hatcher, who were owned by John R. Hatcher of Bedford County, Virginia, before moving to Williamson County. The father of seven sons, Meredith Hatcher, was a knowledgeable farmer who focused on adapting to market conditions and sustainable farming practices. He and his family understood the importance of diversifying and had a variety of crops and livestock, including milk cows, beef cattle, goats, swine, chickens, tobacco, cotton, corn, and sorghum. Taking advantage of new information and farming techniques that were available in the first years of the twentieth century, he maintained and improved his land even as he expanded his acreage. The Hatcher farm was a "gathering place and anchor for the community" because of its stability during hard times and the family's hospitality. Meredith Hatcher died in 1945, and the land passed to three of his sons. Marvin, Sr. received 208 acres, and his brothers, Jimmie and Charlie, received a portion.

Historian Thelma Battle included this photograph of the Hatcher family in her 2023 exhibit, The Hatcher Family: Roots from Owen Hill Road, at Williamson County Public Library. Marvin P. Hatcher Sr. is with his wife, "Mama Sadie," and children, Jessie Mary, Jimmie, Laeunia (standing), and Winnie.

Marvin Hatcher, Sr., and his wife, Era Emma "Sadie" Odell Kinnard Hatcher, were the parents of thirteen children. Marvin served in World War I in the 6th Pioneer Infantry Regiment. Having learned well from his father, he continued to diversify farm products, adding wheat and hay to the variety of row crops and livestock. He was responsible for planting an orchard of peach, apple, plum, and cherry trees. The goal was to be self-sustaining and to be flexible to meet the demands of a changing market. The family worked hard during the week but also hosted regular Sunday dinners, welcoming family, neighbors, and friends. They were also known for inviting visiting ministers who traveled in the area and served the needs of local congregations. The family recalls that Marvin and Sadie saw this as part of their Christian responsibility and as a way to "strengthen the moral and social fabric of rural life."

Jasper Hatcher, Sr., the son of Marvin and Sadie, always had a strong interest in the farm. During his lifetime, his father deeded 61 acres to him. Upon his death, he requested that the remaining heirs sell their land to Jasper to preserve the family's heritage and farm operation. Jasper Hatcher, Sr., took his role as keeper of the family's agricultural traditions as seriously as he did his ministry. A respected clergyman, Elder Hatcher "blended land cultivation with community service." He and his wife, Thelma Thressa Hatcher, and their eight children worked to maintain the farm through the variables of agriculture and the turbulent social times of the 1960s into the twenty-first century. Elder Hatcher successfully combined faith leadership and stewardship of the land, and his contributions as a positive influence in the area and state were recognized. In 1982, he was appointed Colonel Aide de Camp, Governor's Staff; he was named an honorary Member of the Lieutenant Governor's Staff by Speaker of the Senate and Lieutenant Governor John S. Wilder; and he was named an Honorary Member of the Tennessee House of Representatives by Speaker of the House, Ned Ray McWherter. Because of the respect he earned throughout Tennessee, several miles of Highway 96 through Arrington have officially been named the Jasper G. Hatcher Memorial Highway. Elder Jasper G. Hatcher, Sr., died in 2020 at the age of 92.

Today, the children of Jasper, Sr. and Thelma Hatcher own the land. Jasper Hatcher, Jr., is the farm manager and is actively involved in the daily work of the farm along with Lawrence Buford Hatcher, Sr. Products include beef cattle, fruits, and a variety of garden vegetables. The family advises that the house built in 1932 by Marvin and Sadie, and a home built in 1959, remain on the farm, as do two log cabins and outbuildings from the early 1900s. The Hatcher family story is a remarkable one of faith, perseverance, and service to community, county, state, and nation.

The smokehouse and log corn crib date back to the early 1900s, when Meredith Hatcher began farming on his own land.

Sullivan-Givens Farm, 1904

Owen "Tee" and Matilda Sullivan founded a farm in Fairview in 1904.

The Sullivan family is one that helped to establish Fairview and other communities in western Williamson County. Owen Thomas "Tee" Sullivan, born in Dickson County, and his wife, Matilda Jane Tidwell, a native of Williamson County, are the founders of this farm. Tee Sullivan's father was an Irishman who took care of mules that pulled the cannon wagons for the Confederate Army. The family recalls he was still training mules at the age of 90.

When the Sullivans acquired their acreage, this part of western Williamson County, near the Dickson County line, was a region of rather isolated farms. The area was difficult to navigate due to the high hills surrounding the Central Basin, which includes Nashville and Franklin. The old Nashville-Centerville Road ran through the farm, and parts of the route can still be seen. Today, much of the farm can be viewed from U.S. Highway 100, which, before I-40, was the major east-west highway from Nashville to Memphis and goes directly through Fairview.

"Tee" and Matilda Sullivan had nine children, and on their 48-acre farm, they produced turnips, sweet potatoes, Irish potatoes, fruit, swine, and beef cattle. Sullivan used mules to build a stock pond that is still used today. Because of its location on the main road, the farm had a general store beginning in the 1920s. The store building, still standing, was also a bus stop for the Ladd bus line that ran from Nashville to Centerville in neighboring Hickman County.

Ora Sullivan and Kern E. Givens. This couple gave the Sullivan-Givens Century Farm its name.

The next generation of owners was Kern E. Givens and his wife, Ora Sullivan, the daughter of "Tee" and Matilda. With their marriage comes the joining of the two families that give the Sullivan-Givens Century Farm its name. Ora was a schoolteacher for many years and a member of the Home Demonstration Club. Her husband was elected a Magistrate for the 1st District in 1957, and the family recalls that many couples were married on the farm. Their children were Grady McClelland Givens and Wilma Dean Givens (Potchad). Grady, a lifelong resident of Fairview, was the owner of Broken Bow Construction Company.

Grady B. Givens, son of Grady McClelland Givens, along with his wife, Stacey Williams Givens, became the fourth-generation owners. They lived and worked the farm together until he died in 2012 at the age of 40. Stacey and their son, Grady, have operated the farm since that time through the Grady B. and Stacey L. Givens Family Trust.

Stacey placed the land in the Grassland Reserve Program in 2013. This is a permanent conservation easement that will protect the farm from future development other than agriculture, and the acreage will remain in grassland in perpetuity. Timber and fescue are primary crops. An icehouse, fertilizer shed, and equipment shed, all from the 1930s and 1940s, remain part of the daily life of farm work. A barn built by Kern in 1959, a pole barn erected in 1913, and a recent Quonset hut storage barn are all included in *Barns of Williamson County* (Williamson County Historical Society, 2019). Stacey's love for this farm is evident in her documentation of the history of the people and the place, as well as in the protective measures she has taken to preserve the land. She plans for the farm to pass to Grady Givens, the fifth generation, who lives on the farm founded by his ancestors.

1. Few travelers on Highway 100 through the Fairview area know that this building on the Sullivan-Givens Farm was a general store more than a hundred years ago.
2. The barn built by Kern E. Givens in 1959 is one of several outbuildings constructed over the years to support changing farm work.
3. Stacey Givens and a young Grady display their door prize at the Century Farms dinner of the Williamson County Fair. Grady, now 20, helps with the management of the farm.
4. Founder, Owen "Tee" Sullivan, used mules to carve out the stock pond that remains in use today.

Peaceful Valley, 1905

The Peaceful Valley Farmhouse is familiar to those who travel Giles Hill Road in the Flat Creek Community, in the Southeastern part of Williamson County.

Ennis Core Wallace, Sr., who owned Peaceful Valley Farm for more than half a century, was an advocate for the history and community of Flat Creek. He was a farmer, operated a business, and was the co-author of *Flat Creek: Its Land and People*. He and his wife, Allean Harper Wallace, were exceedingly proud of their heritage and worked together and with others to preserve the character of their part of the county while also improving it in many ways.

In 1955, Wallace acquired the farm that he and Allean called home for 59 years. Wallace was related to the farm's founder, C.M. Smithson, through his grandmother. Smithson, a widower, purchased 100 acres in 1905 and, with his four children, raised tobacco, wheat, corn, hay, and livestock. C.T. Wallace, a cousin of Smithson, became the owner in 1925, and Ennis C. Wallace, Sr. acquired the property thirty years later.

The Wallaces and their sons, Ennis, Jr., and Kenneth, raised tobacco, swine, and cattle. They also operated a dairy, and the milking parlor and loafing barn remain. Ennis, Jr., and Kenneth were both officers in the Future Farmers of America (FFA) and both earned the organization's state farmer degrees. Their parents were recognized for their successful farming efforts and were awarded honorary farmer degrees.

Ennis C. Wallace, Sr., and Allean Harper Wallace were sometimes affectionately referred to as the "Monarch" and "Queen" of Flat Creek.

Ennis Sr. is remembered by many farmers as the owner of 4-Star Inc. Farm Equipment in Triune. He was a Past Master and a fifty-year member of Owen Hill Masonic Lodge and an officer in the Flat Creek Community Club for 63 years. Allean worked at College Grove and Page Schools and was a longtime and active member of the Flat Creek Community Club and their church, Edwards Grove. The Wallaces enjoyed hosting family and friends, and they participated in the activities of the children, grandchildren, and great-grandchildren, including livestock shows and ball games. While many of the family members have moved away from Flat Creek, their granddaughter, Lindsey Goolsby, and her daughter, Harper, live on 15 acres that was part of the original Peaceful Valley acreage. Her land remains part of the Century Farm, which requires 10 acres of the original property to remain certified. In recent years, the land was leased to neighboring farmers who primarily grow hay.

Ennis Wallace, Sr., died in 2014, and Allean and Ennis, Jr. became the owners. Allean remained in her home until she died in 2025. The many generous contributions of time, talent, and hospitality of Ennis and Allean Wallace continue to be recognized, appreciated, and remembered.

The milking parlor and multi-use barn remain from the time that the Wallace family operated a dairy.

Luster Farm, 1906

Grant Luster, Jr., and his wife, Mattie Smithson Luster, lived on and farmed their land for over 60 years.

Freed slaves and their descendants found that acquiring property was a challenging, if not impossible, proposition in the years following the Civil War. Generations of slaves and later Black tenants and sharecroppers had enviable experience and knowledge of livestock and crops, but spent their lives building successful farms for others. Many legal and societal obstacles prevented freed men and women and their children and grandchildren from acquiring their own land.

A notable exception is the Luster family. Nelson Luster was born a slave in 1834 to parents who were also enslaved. Nelson and his wife, Betsey, were living in the 21st District of Williamson County with four children in 1870. They were part of the generation that moved from bondage to either tenancy or sharecropping, which was often little better than slavery. The Lusters were unable to achieve their dream of having their own farm, but they helped make that dream a reality for their second son, Grant Luster, Sr., who was born in May of 1864.

Anthony Luster, the fourth-generation owner of the family farm, has worked with cattle most of his life.

In November of 1906, Grant Luster Sr. purchased a farm of just over 80 acres southeast of Franklin on Arno Road. The father of three children; his first wife was named Anna, and he married Sallie Jones in 1920. The family engaged in general farming for their use and to market. Grant Luster, Jr. acquired the farm in 1931 from his father's widow, Sallie Jones Luster. He and his first wife, Nellie, were the parents of James, William, and Nelson, and the farm supported vegetables and livestock. Grant, Jr., married Mattie Jane Smithson in 1933, and they remained on the farm until they died in 1981 and 1991, respectively. Throughout the twentieth century, the Lusters continued to engage in general farming, including corn, wheat, tobacco, sorghum, hay, goats, poultry, swine, dairy cattle, vegetables, and fruits.

Nelson Luster II, grandson of the founders, acquired the property in 1991. Anthony W. Luster, son of Nelson and Cynthia Luster, lived on the farm and managed a beef cattle operation with his father. When I-840 was built, it reduced the farm to just over 60 acres.

In 2022, with so much development surrounding I-840 and the south end of the county, the Luster Farm was in danger of being sold. Anthony, his daughters, and members of the community, however, rallied to support the historic farm remaining in the family. Anthony's daughter, Kendra, and her sisters, Jennifer Luster, Amber Luster, Erica Luster-Smith, and Tasha Luster-Vinson, drafted and began to circulate a petition because they believe the farm means so much more than the value of the land. The petition, eventually signed by thousands of people, stated:

"The farm represents the dream of freed slaves who yearned to own their own land, and provide a life-changing legacy for their descendants."

Kendra and her sisters wanted people to know that Black farmers and this farm are important now, just as they were in the past. In doing so, she mentioned the small family cemetery on the farm, where many of her ancestors who worked so hard to make a life for themselves and those who followed them are buried.

Kendra explained, "All we ask is that we can continue to keep it in the family and pass it on for more and more generations to come!" The successful outcome of this effort was that Anthony Luster is now the sole owner of the farm, and his daughters represent the fifth generation. Luster observes:

"Families of the past, while working with hand tools and beasts of burden, seemed closer knit than perhaps today, but the legacy, the heritage, and the bond of the land still continues."

The family gathers for celebrations whenever they can. Kneeling is Kendra, and on the second row are Valerie, Erica, Jordan, Tasha, Kynnedi, Amber, with Anthony and Jennifer in the back.

Children and grandchildren representing the fifth and sixth generations with patriarch Anthony are: first row, Desarae; second row, Erica, Caden, and Jennifer; third row, Kendra, Anthony, and Amber.

Jennifer Luster, daughter of Anthony, maneuvers the Massey Ferguson tractor on her ancestor's farm.

Bledsoe-Sullivan Farm, 1906

The Bledsoe family, including the matriarch Jean Sullivan Bledsoe, received their sign and certification in 2017.

The twentieth century had hardly begun when Zachariah Joseph Sullivan and his wife Parmelia Caldonia "Callie" Flowers Sullivan registered the deed for land in western Williamson County near the Hickman County line on November 3, 1906. The Sullivans were the parents of nine children, and the family raised corn, wheat, livestock, chickens, and vegetables.

Nearly a half-century later, Imogene "Jean" Sullivan, a daughter born in 1928, acquired the property on March 17, 1951. "Grandmother Jean" was well-known in the Fairview community for her baking and cooking, which she readily shared with others;

her quilting skills; her presence at church services and gatherings; and her support at sporting events for her children and grandchildren.

Ricky Bledsoe farmed the family property for most of his life and was a coach for 26 years.

Willis Bledsoe, born in 1925, was a veteran of the U.S. Marine Corps who served in the Pacific Theater during WWII and was awarded a Purple Heart. After returning to Williamson County, he and Jean married in 1949 and settled on a portion of her family's land. Willis farmed and worked in construction. They and their children — Wanda (Welch), Ricky, and John — had row crops and raised livestock. Family and friends remember Willis as an optimistic man who encouraged them and made others believe in themselves, while also offering witty sayings. Both Jean and Willis were active members of New Hope Baptist Church in Bon Aqua and lived on the farm until they died in 2014 and 2017, respectively. Ricky Bledsoe and his wife, Donna Nutt, acquired 137 acres of the family farm in 2006. Both Ricky and Donna taught in Williamson County schools while farming and raising Kori and Drew. Donna, also a Fairview native, was an outstanding basketball player at Belmont University and taught Physical Education and coached in local schools. Ricky coached for 26 years and served as a mentor to many athletes and students. Donna and Ricky were married for 32 years at the time of her death in 2013. Ricky continued farming with Drew, growing hay and cattle on the acreage which was certified as a Century Farm in 2017. Ricky passed away in 2023, leaving a legacy of family farming, teaching, and community service.

This cemetery has long been the burial site for members of the Sullivan and Bledsoe families including Callie Flowers Sullivan, her daughter Imogene Sullivan Bledsoe, and third generation farm owners, Ricky and his wife Donna Nutt Bledsoe.

Nichols Farm, 1909

The farmhouse dates to about 1803, at least a century before the McFarlins purchased the land.

Williamson County was a major contributor to Tennessee's dairy industry, and milking parlors and dairy cows were part of the landscape. The income from milk was crucial to farm families, and the economic importance and stability of the dairy industry, which included dairy processing plants such as Purity, Meadow Gold, and Mayfield (a Century Farm in McMinn County), were significant. As development spread into the county, along with inconsistent milk prices and a shortage of laborers needed to milk twice each day and maintain the herds, the number of dairy farms decreased until few remain today. The Nichols Jersey Farm, as it was known for years, was one of the survivors until 2018. The traffic on Kidd Road in Nolensville made it unsafe to drive the herd across to graze in pasture, and, with fewer dairies, it was hard to get anyone to haul the milk to processing plants. The family decided to switch to raising Black

Angus cattle, growing hay, and planting a large garden for their use and for market. Farm families are, and must be, adept at transitioning so they can continue farming. The Nichols family understands this concept very well.

Benajah Hill McFarlin

Mary J. Turner McFarlin

The farm was founded in 1909 by Benajah Hill McFarlin and Mary Jane Turner McFarlin who married in 1880. They had eight children. Primary crops were corn, wheat, and oats. The family ran a general store at the corner of Kidd and Nolensville Roads until the 1940s and operated a grist mill and sawmill. Current residents of the fast-growing area may be interested to know that Mill Creek was named after the McFarlin mills.

In 1923, Lena Senethius McFarlin became the second-generation owner of the farm. Married to Berry D. Nichols, they had children, Mary Jane, Sue Mildred, Rebecca, Robert, Douglas, and Herbert. The family raised row crops and began the dairy operation that lasted nearly a century. The hay and loafing barn was built in the 1940s by Will Fly after an earlier barn was struck by lightning and burned. Herbert Nichols became the farm's owner in 1972 and built the current milking parlor in 1981.

Herbert and his wife, Agnes Jenkins Nichols, live in a log house that is believed to have been built around 1803. The original dwelling had two pens or rooms downstairs and two upstairs, and was added to in 1935. In the 1940s, Emmitt Jenkins, Agnes's father, tore down the old store and built a home for his family from the lumber. Herbert and his son Mark worked the land together and raised dairy cattle and hay for years.

Siblings Mark and Leigh Ann Nichols showed Jersey cattle and won numerous awards. Mark and his wife, Sandra, raised their three children, Brandon, Kyle, and Brittany, on the farm. All were involved in 4-H. Today, Mark and Sandra are the farm's managers, and Kyle, Brandon, and their families help as needed.

The Nichols are acutely aware of, and are reminded daily of, the changes and challenges facing farmers in the fast-growing community. Mark, a life-long resident of the farm, explains, "Gone are the days of a one-lane gravel road going through our farm where you rarely saw a car. Our main challenge today is the steady stream of traffic all day long, sometimes with backups that extend from one end of the farm to the other. We must strategically plan our work around the barn based on traffic. It is challenging and dangerous to move our hay equipment in the summer and navigate our cattle trailer to load cattle."

Nevertheless, Sandra reflects that "It has been the joy of our lives to raise our children here on our family farm and teach them the value of hard work, especially when showing our registered Jerseys at county shows and the TN State Fair. They have learned so many life lessons." As to the future of the Nichols Farm, it is hard to predict. With development surrounding the farm and increasing traffic issues, the Nichols family "isn't quite sure how long we can continue to farm." Their plight faces most farming families in Williamson County.

Brandon, Mark, and Herbert Nichols operated the dairy until 2018. The farm and family have been featured in several publications.

NICHOLS FARM, 1909

Four generations of the Nichols family gathered at the homeplace to celebrate Christmas.

Mark Nichols offers farm fresh vegetables.

The Nichols Farm has an extensive vegetable garden.

Top Left: Mark Nichols with his prize Jersey, Dreamer Ann, who was named Supreme Grand champion of the Mid-South Fair in Memphis in 1985. *Top Right:* Mark's daughter, Brittany, and her Jersey, entered shows, as did her brothers when they were in 4-H. *Bottom:* The next generation of the family learning about farming includes Landon and Kendall (left) and Myla and Emmitt (right).

Penn Hollow Farm, 1913

The 1876 farmhouse is home to the current owners who appreciate its history.

As the twentieth century progressed, farms previously owned and worked by individuals and families who had settled throughout the nineteenth century came on the market. James Carol Sullivan and his wife, Lillie A. Tidwell, purchased two tracts of approximately 150 acres located five miles northwest of Fairview in 1913. In deference to Robert Penn, who owned the land and built a house on it in 1876, the nation's centennial year, this Century Farm is named Penn Hollow.

The Sullivans operated a steam-powered sawmill and grist mill on the farm. Additionally, they had a store, so people living nearby were accustomed to coming to this farm to purchase supplies, tools, and lumber while also having their wheat, oats, and corn ground.

By 1926, Nancy Jane Sullivan, sister of James Carol Sullivan, and her husband, W. J. Fisher, had acquired the farm. They continued to operate the mills and store and raised cattle, hay, and corn. In 1943, G. B. Fisher, nephew of James Carol and son of Nancy Jane and W.J., acquired 142 acres. He and Elvana Anderson Fisher had four children, and the family raised tobacco, tomatoes, corn, hay, cattle, and swine. A parcel of land was donated in 1950 by the Fishers for the Liberty Hill Church of Christ, which remains a large and active church on Old Cox Pike.

Helen Ann Clark is the great niece of the founders. She and her husband, William P. Clark, acquired 35 acres of the original farm in 1991 and are the managers. They live in the original house built by Robert Penn, although J.C. Sullivan rebuilt parts following a fire, and it has, of course, undergone remodeling over the years. The mill house was destroyed by a storm in 2021, and the store was torn down in 2022 because it was unsafe and no longer used. Penn Hollow was certified as a Century Farm in 2023, having been in the family for 110 years.

Several children of Nancy Jane Sullivan, the second owner and sister of the original owner, gathered in front of the farmhouse in September of 1939.

Top: Perkins Branch remains a vital water source on the farm. It is named after the original owner of the Tennessee Land Grant, Ebbin Perkins. Perkins Branch runs into Big Turnbull Creek on its way to the Harpeth, then the Cumberland River. **Middle and Bottom:** *A good cutting of hay ensures the cattle will have feed all winter.*

Willow Creek, 1915

Virginia Pearre Egbert with her husband, Paul Egbert, and their daughters, Rose Elaine and Camille, in front of one of the tobacco barns built by Virginia's grandfather, her father, and uncles.

As current generations assume ownership and management of historic family farms, they are influenced and rely on the wisdom of previous generations. They are also grateful for existing barns, storage sheds, corn cribs, and other buildings that continue to support the farm's business. Willow Creek, also known as the Pearre Farm, is in Waddell Hollow. In 2010, the Pearre family, at that time headed by patriarch Paul H. Pearre, decided to place their acreage into a trust. Paul, along with his son Joseph Pearre, his daughter Virginia Pearre Egbert, her husband Paul Egbert, and Jimmy Thornton, a nephew, became the owners.

Paul lived most of his 92 years as a farmer on the land his parents, Joe, Sr., and Eunice Pearre, purchased in 1915 for $10,000 from the Hughes family. Joe, Eunice, and their children, Robert, Buford, Virginia, Milton, Paul, and Edith, operated a diverse farm and raised tobacco, corn, wheat, oats, and barley, and had sheep, dairy and beef cattle, swine, and poultry, along with the essential large vegetable garden. Joe, Sr., along with his sons, built two tobacco barns that remain, as well as a mule barn rebuilt after the original one was destroyed in a storm. A smokehouse, potato house, and blacksmith shop still stand, indicating how self-sufficient this farm was for decades. Buford died in 1961, and the farm passed to the three surviving sons and to Eunice in 1977 at Joe Sr.'s death.

For over 50 years, Paul and his older brother, Milton, managed and farmed 400 acres, expanding beef cattle and hay production. Jimmy Thornton, grandson of the founders, was also involved in the farm work until his death in 2022. During this half-century when the Pearre brothers owned the farm, tobacco was the major cash crop. When Paul was 59, he married Rose Ann Nix, and they raised their two children, Joseph and Virginia, on the family farm. Rose Ann was involved in Girl Scouts for twelve years, and both she and Paul were regular supporters of their children's school events and community activities. Willow Creek was the name Paul and Rose Ann chose for the farm.

Paul and Milton Pearre, their brother, Robert, and their mother, Eunice Pearre.

Paul and Rose Ann Nix Pearre, the second generation to live and work on the farm, with their children, Virginia and Joe.

Paul's family described him as a "man of compassion, friendship, and conviction, which showed in everything he did in life, whether it was being a farmer, friend, or father." As a youngster, he plowed with a team of mules and had his own tobacco patch. Paul's only time living off the farm was during his service in the post-WWII occupation of Germany. Pearre Creek Elementary School, which opened in 2010, was named after the family. Milton, who died in 2011, and Paul, who died in 2019, were extremely proud of this honor.

Siblings Virginia Pearre Egbert and Joseph Hunter Pearre are the owners and operators of the farm, along with Virginia's husband, Paul Martin Egbert. The Egberts and their two daughters, Rose Elaine and Camille Pearre, enjoy living in a new house they built on Willow Creek Farm in 2018. They incorporated materials from the farm in their house, including quarried yellow limestone for the fireplace. For their kitchen floor, they used seasoned wood that was harvested from timber on the property and milled at Fox's sawmill some years ago. From an old tobacco barn, they salvaged rare chestnut wood for a bathroom vanity.

The farm currently supports beef cattle and hay to feed the herd, with surplus sold to local farmers. The current owners intend to continue the family trust and pass the farm to the next generation. Virginia reflects on her family farm, stating, "I am the third generation of Pearres to live on this beautiful farm. My father continued the tradition and worked to ensure that my brother and I could do the same. I am blessed to have the same opportunity to raise the 4th generation on the land. While I didn't have the chance to meet many of my family members and relatives, including my grandparents, I feel connected to them and their legacy through this farm."

> **"My grandfather purchased our farm in 1915 in hopes of raising his family and providing a home for generations to come."**

Joe Pearre, Paul Pearre with granddaughter Rose Elaine Egbert, Virginia Pearre Egbert, and Paul Egbert gather for a treasured three-generation photograph.

WILLOW CREEK, 1915

The compact corn crib

The main livestock and hay barn

Thomas Jefferson Byrd Farm, 1920

Thomas J. Byrd, Jr., owned the farm from 1965-2011.

Johnson Hollow Road, though part of fast-growing Burwood, maintains its rural character from a century ago, even though it is a short distance from I-840. The Byrd and Johnson families have been deeply rooted in the community for decades.

The two families were joined when Thomas Jefferson Byrd, Sr. of Maury County, wed Rachel Kathryn "Katie" Johnson in 1899. The couple acquired 86 acres just into Williamson County and began farming there in 1920. They eventually had ten children, though not all survived childhood.

The house built for the large family in 1924 still stands. Tobacco, dairy cattle, hay, and vegetables were some of the farm's products. Thomas Jefferson Byrd, Jr., and Betty Plant Byrd were married in 1947 and became the owners of the land in 1965.

While continuing to own and manage the farm, Tom and Betty lived in Nashville, where he worked as a supervisor at Nashville Electric Service. This would account for the change from dairy cattle to beef cattle, though hay and tobacco were still raised.

After his father died in 2011, Phillip W. Byrd, Sr., acquired 25 acres of farmland owned by his grandparents. Married to Debra Lynn Byrd, he manages the farm and raises hay. The current owners are Phillip and Debra, as well as Lynn, Phillip Jr., and Hannah Byrd. The farm was certified as a Century Farm in 2023.

THOMAS JEFFERSON BYRD FARM, 1920

The main dwelling on the Byrd farm was built in 1924 and is now over 100 years old. It remains on the farm.

Crestview, 1922

An aerial view of Crestview is a reminder of the remarkably scenic landscape that drew early settlers and subsequent farm owners to this part of the county. This farm is the recipient of agricultural awards for best farming practices by its current residents, the Shirling and Polk families.

In 2001, Mary Louise Osburn Stallings made the significant and timely decision to place her family farm under a conservation easement with the Land Trust for Tennessee to prevent the development of 112 acres in one of the areas of Williamson County that was beginning to experience rapid growth. Her decision was prescient given the changes in just twenty-five years. For more than a century, Crestview has maintained a history of progressive farming and land stewardship in the Arrington community.

This parcel became the property of Leslie Wilson Osburn and Mary McMurray Osburn in 1922. Even as the country and the state were in economic upheaval, the couple managed to raise the substantial sum of $3,200 to purchase the property on the courthouse steps. The Osburns began raising beef cattle, tobacco, poultry, hay, and fruits for their family and to market. They operated a dairy for decades and, by 1952, sold 170 gallons of whole milk daily. Three barns and the farmhouse date from the 1920s and 1930s and have been updated and maintained. Their children, Mary Louise and Mack, grew up working on the farm.

The gambrel roof barn at Crestview is a focal point of the farm.

Mary Louise Osburn was a 1939 graduate of Franklin High School and served in the Women's Army Corps (WAC) during World War II. She and her husband, Virgil Stallings, acquired the property in 1972. Their children are Leslie Osburn Stallings and the current owner, Virginia "Ginger" Dallis Stallings Shirling. After retiring, Mary Louise returned to the farm and made many improvements in addition to preserving the land through the conservation easement. In 1995, the farm was recognized in the United States Department of Agriculture (USDA) Grasslands Reserve Program.

Ginger and her husband, Milo Rex Shirling, Jr., live on the farm with their daughter Dallis Shirling Polk, her husband Travis, and their daughters Josephine and Eleanor. Cattle, horses, and hay are raised on the property. The three generations participate in pastureland conservation and other programs of the National Resources Conservation Service and Farm Service Agency of the USDA. In 2011, the farm received the Harpeth River Steward Award and the Agriculture Award by the Harpeth Conservancy for the family's efforts in best farming practices.

FORMER CENTURY FARMS REMEMBERED

It is appropriate to pay homage to farms that were once certified as Century Farms but no longer qualify for that designation. The families that owned and farmed these places contributed to the county in many ways and deserve recognition and gratitude.

As part of the 1975 policy of the Tennessee Department of Agriculture that established the Century Farms Program, a property ceases to be recognized as a certified Century Farm if:

> *(a) The farm no longer has ten acres of the original farm;*
> *(b) No owner of the farm is a resident of Tennessee;*
> *(c) The land does not produce $1000 annually in agricultural income; or*
> *(d) The property has been sold out of the family.*

These guidelines have governed the program for fifty years.

Several of the farms included in this section were among the first certified Tennessee Century Farms submitted from Williamson County in the early years of the statewide program. Most were included in *Tennessee Agriculture: A Century Farms Perspective*, Carroll Van West, Tennessee Department of Agriculture, published in 1986. A few are now repurposed green spaces; on others, current owners have retained buildings and produce some agricultural products; one farm was placed in a conservation easement to preserve the landscape even though sold out of the family; but most have been sold and developed for residential or multi-purpose uses as growth continues unabated across the county.

FORMER CENTURY FARMS REMEMBERED

Moss Side Farm, 1810

Rivers Meet Farm, 1816

Montrose Farm, early 1800s

Midway Farm, 1832

Hillsdale Farm, 1842

Mockingbird Hill Farm, 1852

Rodgerswood, 1853

Woodland Farm, 1857

Creekside, 1860s

Reynolds' Grant Farm, 1865

Dripping Springs Stock Farm, 1869

Carter's Acres Farm, 1870s

Westbrook (Short Farm), 1887

Cedar Lane Farm 1, 1896 – Cedar Lane Farm 2, 1897

Cartwright Farm, 1898

J.R. Givens, LLC, 1902

Moss Side Farm, 1810

The family of Evie Moss and Alphonso Gibbs, and a hired hand, gathered in front of a splendid automobile, perhaps their first, which was the reason for the photograph.

Established by Francis Giddens, a Revolutionary War veteran from Virginia, the farm was just west of Thompson's Station. Subsequent family owners were Sarah Giddens Moss, Evie Moss, and her husband Alphonso Gibbs, who served in the Tennessee House of Representatives and the Williamson County Court for 20 years. In 1976, when the farm was certified as a Century Farm, it was owned by Malcolm Moss Gibbs, who was raising cattle and tobacco. In the 1980s, an 1814 house and other outbuildings remained.

Rivers Meet Farm, 1816

Near Leiper's Fork on Southall Road, Rivers Meet was a 2000-acre plantation founded by Henry Hunter. An impressive brick dwelling, a smokehouse, a well house, and a barn were still intact when Harriett Hunter McCullough owned the property, which was operated and worked by Harry Sanders in the 1980s. The land was later divided into tracts and sold. The Italianate residence, built in 1875, is listed in the National Register of Historic Places, and the current owner retains about 13 acres around it. Also on site is a log slave dwelling that later became a tenant house and, for several years, was an antique shop. Nearby is the Leiper's Fork Distillery, which occupies a parcel of the original Hunter farm.

Top Left: Harriett Hunter McCullough and Robbie Hunter across Southall Road with the house in the background. *Bottom Left:* Once a slave cabin and later a tenant house, this log building predates the main dwelling. *Right:* Rivers Meet is one of the county's most recognized historic homes.

Montrose Farm, early 1800s
Midway Farm, 1832

Montrose Farm, early 1800s

Associated with the Sprott and Grigsby families, this farm was located twelve miles south of Franklin. The family donated a portion of their farm's acreage to the Bethesda United Methodist Church, which dates back to 1823, and the parsonage. In the 1980s, Cleo Grigsby owned the farm, which the Bond brothers worked.

Midway Farm, 1832

Traveling on Hwy. 31 from Franklin to Brentwood, one of the most impressive properties was Midway Farm. Owned by Lysander and Elizabeth McGavock, it was one mile south of Brentwood. McGavock owned over 1200 acres by 1855. He left the plantation to his four daughters. The farm was partitioned in the 1950s and is the site of the Brentwood Country Club and Golf Course, which retained the two-story brick dwelling of the McGavock family.

Hillsdale Farm, 1842

This farm was in the Meacham family for generations, and each owner expanded and added acreage. By the 1980s, Harold Meacham owned one of the largest farms in the county. The farm was divided and sold over the years.

Top: The family of M.A. and Alice Kirby Meacham of Garrison gathered for their 50th anniversary. Bottom: Hillsdale was on either side of Garrison Road. Many of the barns and outbuildings remain on this parcel and other acreage that were once part of a major agricultural enterprise.

Rodgerswood, 1853

In 1853, Roche Carter Rodgers received the farm in the Bingham community from her brother, John C. Carter. She was also a sister to Fountain Branch Carter, owner of the Carter House in Franklin. When the Century Farms Program was announced in 1975, then-owner Nan Chapman Rodgers was among the first in the county to certify her farm. She received her sign in 1977-78 at the Tennessee State Fair and proudly displayed it at her home until she died in 1979. The farm was later sold.

Top: To those who knew her, Nan Chapman Rodgers was a beautiful woman, no matter what age. Bottom: The once comfortable and beautifully maintained gable-front and wing house was home for many years to Nan Chapman Rodgers. The photograph shows its condition before the farm was sold.

Woodland Farm, 1857

Located ten miles northwest of Franklin on the old Natchez Trace, this farm was the home of the Moran family. It was notable for its nineteenth-century log dwelling and outbuildings. The second generation, headed by James Moran, was a proponent of progressive farming, and under his management, Woodland became a successful operation. The Great Depression, however, nearly brought Woodland to an end, for James had invested heavily in the National Bank of Franklin which collapsed. The family worked hard and led a conservative lifestyle to restore the farm to production. Ann Moran inherited 195 acres in 1973 and supervised the work of her brother-in-law, Paul Kinnie, who raised tobacco and cattle. After Ann's death, John Kinnie, Paul's son, served as caretaker until the livestock was removed and the house and outbuildings came down after the farm was sold.

Margaret Fain and Samuel Houston Moran were the founders of the farm, and Ann Moran was the last member of the family to live in the unique log dwelling, which was two log houses with an enclosed dog trot. It was listed in the National Register of Historic Places. The house was dismantled in 2022 when the farm was sold.

Mockingbird Hill Farm, 1852
Reynolds' Grant Farm, 1865

Mockingbird Hill Farm, 1852

Related to the family at Moss Side, William and Sarah Giddens Moss purchased 174 acres and later added nearly 200 more acres. Grains, cotton, and livestock were raised on the farm. William and Sarah helped to organize the Thompson Station Methodist Church. They also freed their slaves before the Civil War. By the 1980s, Mrs. Kennedy Moss (Elizabeth) Gibbs was supervising the production of cattle, hay, and tobacco with sharecropper Harvey North.

Reynolds' Grant Farm, 1865

Richard and Susie Scales Reynolds established their farm of 110 acres in the southeastern part of the county at the close of the Civil War. They were the parents of eleven children and attended Edwards Grove Methodist Church. Reuben Scales and his wife, Effie Hargrove, acquired the farm in 1899 and raised a variety of products for table and market. The third-generation owner was James King Reynolds, who acquired the farm in 1929. He managed the farm through the Great Depression and for more than 50 years. He was the owner and operator in 1976 when the farm was certified as a Century Farm.

Creekside, 1860s

The 1835 brick I-House on Hwy. 31 North will be protected by its new owner—the City of Franklin.

Long a landmark north of Franklin on Hwy. 31 (now at the corner of Mack Hatcher Parkway), Creekside, listed in the National Register of Historic Places, was owned as early as the 1860s by John B. and Cynthia McEwen. The farm was given to their daughter, Florence, and her husband, Rev. William Rosser. The two-story brick house and spring house, however, date to 1835. The farm remained in the family as an active farm until it was purchased from the heirs by Franklin Preservation Partners in 2023 with the intention of saving it from residential development. The *Williamson Herald* of July 17, 2025, reported the final purchase of the 61.8-acre Creekside property by the City of Franklin. A master plan for the property is being prepared, which offers several options and will protect the land and the antebellum house.

Dripping Springs Stock Farm, 1869

Soon after the end of the Civil War, Milton and Pina Jane Meacham purchased 120 acres eight miles west of Franklin. They were the parents of eleven children. In 1903, their son, John, and his wife, Elizabeth, became the owners. The couple were progressive farmers, and John practiced improved agricultural methods and added sheep and tobacco to their products. Elizabeth was active in the newly formed Bingham Home Demonstration Club. Two of their children, Margherita Meacham and Florence Meacham Pewitt, owned the farm in the 1980s. Margherita died in 1990, and Florence, the last of the family line, died in 1993.

Top: Sisters Margherita and Florence are not intimidated by a prize bull on the Meacham farm. The dwelling at Dripping Springs Stock Farm includes a two-pen log house updated to a gable front and wing house in the early 20th century. *Bottom:* Elizabeth Meacham with daughters Margherita and Florence.

Carter's Acres Farm, 1870s

Lottie Carter Ashworth was one of the owners of Carter's Acres when it was designated a Century Farm in 1976.

Jeff Carter was born a slave in Williamson County and served as a body servant to his master in the Battle of Chattanooga. He acquired 30 acres, 15 miles southwest of Franklin, on Sugar Ridge in the 1870s, and expanded his farm to 70 acres, on which he operated a diverse operation, including cotton, sorghum, and several types of livestock. Fred Carter purchased some of the shares held by Jeff's heirs and his widow in the 1950s. He and his wife, Alice Watson Carter, also owned a variety of livestock, poultry, and grains. Lottie Carter Ashworth also owned part of the original farm in 1976 when it was designated a Century Farm. Carter's Acres was one of the first African American Century Farms to be certified in the state.

Westbrook, 1887
Cartwright Farm, 1898

Westbrook (Short Farm), 1887

Located just 3 miles west of Franklin on Hwy. 96 is the property that Jesse Armistead Short and Benjamin Franklin Short purchased. On almost 200 acres, Jesse raised barley, wheat, oats, and corn while Benjamin Franklin specialized in livestock. The next owner, Jesse Edlin Short, also raised Persian sheep, which were prized for their black curly wool. Ten acres of the farm were sold for the building of Highway 96 West in the 1960s. About that same time, the Shorts developed a Grade A Dairy, which they continued to operate until 1994. In 2023, the Short family began discussions with the City of Franklin to develop the remaining 200 acres into an "agrihood" or farm-focused community called "Armistead," which will "blend agrarian traditions with modern living."

Cartwright Farm, 1898

Timber, burley tobacco, turnip greens, and hogs were the main products on the farm south of Franklin founded by Benjamin Dotson. He and his wife Ann and two of his brothers, Walter, who was married to Ella Presley, and Sylvester, whose wife was Minnie Pearl Gosey, shared ownership of the farm. The large extended family raised cows, horses, mules, chickens, hogs, and goats, as well as a variety of fruits and vegetables. Walter and Ella, parents of seven children, eventually bought about 168 acres from Benjamin. Their daughter, Carolyn, and her husband, Thomas Burns, acquired the farm in 1961. With their four children, Sara, Carolyn, Angela, and Martha, they raised tobacco, hay, turnip greens, hogs, cows, horses, and mules. By 1980, their daughter Martha owned the farm and lived there with her daughter and grandson.

Cedar Lane Farm I, 1896
Cedar Lane Farm II, 1897

Dorothy McCord Ryan and her daughter, Bettye Cason, enjoyed attending the Century Farms event at the Williamson County Fair.

Annie Lou Reed McCord and Walker L. McCord were the second-generation owners of a farm located in Bethesda. Their daughters were Dorothy McCord Ryan and Elizabeth McCord Crunk. Active operators and managers of their properties each chose to apply for Century Farm certification for the acreage they inherited from their parents. Dorothy was the owner of Cedar Lane Farm I, and Elizabeth was the owner of Cedar Lane Farm II. Elizabeth Crunk managed and worked her family farm and the farm she had lived on with her husband until his death. She decided to place the two farms under her care in a conservation easement with the Land Trust for Tennessee to preserve this green space from development, even if sold out of the family. Elizabeth died in 2013. Cedar Lane Farm I was sold out of the family after the death of Dorothy at age 94 in 2018.

J.R. Givens, LLC, 1902

Located on Caney Fork Road in Fairview, Sidney Pollard Givens and Lucinda Sullivan Givens were the founding couple. Beginning in the early 1950s, when owned by Johnnie Russell "J.R." Givens, the farm had Hereford beef cattle and a few dairy cows. Corn, hay, and vegetables were grown, with tomatoes sold at the Nashville Farmer's Market and H.G. Hills grocery stores in the area. The farm remained with the heirs until it was sold in 2022.

A rare goose house is one of several outbuildings.

The farm on Caney Fork included a substantial livestock barn.

Selected Sources

Books

Tennessee: Its Agriculture and Mineral Wealth, J.B. Killebrew, Travel, Estman, and Howell, 1876.

Tennessee Agriculture: A Century Farms Perspective, Carroll Van West, Tennessee Department of Agriculture, 1986.

Tennessee Encyclopedia of History and Culture, Carroll Van West, Editor in Chief, Tennessee Historical Society, 1998, (and online).

Tennessee Farming, Tennessee Farmers: Antebellum Agriculture in the Upper South. Donald L. Winters, University of Tennessee Press, 1994.

Barns of Tennessee, Caneta Skelley Hankins and Michael Thomas Gavin, Tennessee Electric Cooperative Association, 2009.

Plowshares and Swords, Tennessee Farm Families Tell Civil War Stories, Caneta Skelley Hankins and Michael Thomas Gavin, Center for Historic Preservation at Middle Tennessee State University, 2013.

Up from the Mudsills of Hell: The Farmer's Alliance, Populism, and Progressive Agriculture in Tennessee, 1870-1915, Connie L. Lester, University of Georgia Press, 2006.

Wheat Culture in Tennessee, J.B. Killebrew, The American Printing Company, 1877.

Back Home in Williamson County, Lyn Sullivan Pewitt, Hillsboro Press, Franklin, TN, 1986, reprint 1996.

Barns of Williamson County, Tennessee, Caneta Skelley Hankins and Rick Warwick, Williamson County Historical Society, 2019.

College Grove Williamson County, Tennessee History & Families, History and Genealogy Group, Fifty Forward, College Grove, Panacea Press, Nashville, 2011.

Flat Creek: Its Land and Its People, Compiled and Written by Ennis C. Wallace, Sr., Jo Ann Perry, Marjorie Redmond, Martha Ann Hazelwood, Woodward & Stinson Printing Company, Columbia, TN, 1986.

Historic Williamson County: Old Homes and Sites, Virginia Bowman, Sovran Bank, Franklin, Tennessee, 1971, reprinted, 1989.

National Register Properties of Williamson County, Tennessee, Pearce, Warwick, and Hasselbring, Hillsboro Press, 1995

The following books by Rick Warwick:

Leiper's Fork and Surrounding Communities (1999)

Out There in the First District (2001)

At Home with Working Folks in Williamson County (2018)

Bethesda and Surrounding Communities (2023)

Burwood and Beyond (2024)

"Williamson County", pp. 787-810, 865-1059. In *History of Tennessee*. Nashville, TN: The Goodspeed Publishing Company, 1886.

Articles

Williamson County Historical Society Journals are cited within the text.

"Potash from Pyramids: Reconstruction DeGraffenreid (40WM4) – A Mississippian Mound Complex in Williamson County, Tennessee," Kevin E. Smith, *Tennessee Anthropologist*, 19 (2): 91-113, 1994.

"Archaeology at Old Town (40WM2): A Mississippian Mound Village Center in Williamson County, Tennessee," Kevin E. Smith, Tennessee Anthropologist, 18 (1): 28-44, 1993.

"Williamson County In Black & White", *Williamson County Historical Society Journal*, No. 31, 2000.

Online Sources

Getty Images

Library of Congress

United States Census Records, various years, digital online.

United States Department of Agriculture, Department of Agricultural Statistics, various years, digital online.

United States Department of Agriculture *Circular No. 33*, p. 48, H. H. Bennett. Washington, DC, U.S. Government Printing Office, 1928. Available in digital format from the Library of Congress.

Tennessee State Library and Archives, TEVA digital and online photographs.

The *Review Appeal* and the *Williamson Herald*, specific issues are cited within the text.

Obituaries and the Find-a-Grave website.

Other Sources

Tennessee Century Farm applications, Gore Center, Middle Tennessee State University. These applications, most usually prepared by the families date from 1975 to 1925 and include documents and photographs as well as the generational history and contributions of farm families.

The Warwick and Williamson County Historical Society photograph collection.

Index

11th Cavalry, CSA, 66
21st District, 154
4-H Club, 103, 136, 140, 145, 161
4-Star Farm Equipment, Inc., 153
6th Pioneer Infantry Regiment, 147

A

African American Century Farm, 146, 154, 187
African American farmers, 135
African slaves, 2
Agricultural Stabilization Service, 104
Agriculture
Statistics, 2
Agriculture after the Civil War, 87–89
Agrihood, 188
Agritourism, 76
Allen, Carol
    see Bond, Carol Allen, 98
Allisona, 146
American Polled Hereford Assoc., 60
Anderson, Bess, 140
Anderson, Bettie
    see Bond, Elizabeth J. "Bettie," 139–140
Anderson, Brownie, 140
Anderson, Bud, 139, 140
Anderson, Elvana
    see Fisher, Elvana Anderson, 166
Anderson, J.W., 139
Anderson, James, 140
Anderson, Katie
    see Linton, Katie Anderson, 91
Anderson, Mary Frances, 140
Anderson, Thomas P., 139
Anderson, Tommie, 140
Anderson, Walter, Jr. "Bud," 140
Anderson, Walter, Sr., 139–140
Armistead, 188
Arno, 146
Arrington, 148, 175
Arrington Vineyards, 42
Ashworth, Lottie Carter
    see Carter, Lottie, 187
Aurelia Acres, 42–45

B

Badlands, 104
Bag End Farm, 73–75
Bailey, Evan, 13
Bailey, Lola, 13
Bailey, Misty, 13
Barker, Annie Lou
    see Grigsby, Annie Lou Barker, 120, 127
Barker, Franklin Neil, 120–121
Barker, Franklin Neil "Frank," 121–122
Barker, Franklin Neil, Sr., 121–122
Barker, Hilda Stokes
    see Stokes, Hilda, 121
Barker, Lottie, 127
Barker, Lottie Lee Lavender
    see Lavender, Lottie Lee, 120
Barker, Nancy
    see, Craig, Nancy Barker, 121–122
Barker, Polly
    see Duncan, Polly Barker, 120
Barker, R.H., Jr., 120
Barker, R.H., Sr., 120
Barker, Robbie
    see Norman, Robbie Barker, 120
Barker, Robert H., Sr., 127
Barker, Robert Houston, Sr., 120
Barker, Ruth, 120
Barker, Sandra
    see, Hayboer, Sandra Barker, 121–122
Barker, Talitha
    see Crowson, Telitha Barker, 120
Barker's Hillview Farm, 120–122
barn, rare four-crib Appalachian-style, 117
barn, quonset hut, 150
Battle Ground Academy (BGA), 69, 74, 78, 123–124
Battle of Chattanooga, 187
Battle of Franklin, 69
Battle, Bettie
    see Green, Bettie Battle, 41
Battle, Elizabeth Ogilvie, 22–23
Battle, Rob, 23
Battle, Thelma, 147

Battle, Valerie
	see Kienzle, Valerie Battle, 23
Battle, William Robert, 22
Beasley, Ella
	see Hunt, Ella Beasley, 106–107
Beasley, Evaline, 106–107
Beasley, Newton Cannon, 107
Beasley, Willie
	see Jones, Willie Beasley, 107
Beech Hill, 20–23
	see Ogilvie Farm, 20
Beech Hill Farm, 7
Beech, Harriet
	see Walker, Harriet Beech, 143
Belmont College, 17
Belmont University, 159
Bethesda, 59, 73, 83, 97–98, 123–125
Bethesda Schools, 73, 74, 98, 103, 124
Bethesda United Methodist Church, 180
Bicentennial, national, 3
Big Turnbull Creek, 167
Bingham community, 35, 182
Bingham Home Demonstration Club, 186
Black farmers, 154–157
Blacks after the Civil War, 87–89
blacksmith, 59
Bledsoe, Drew, 159
Bledsoe, Imogene "Jean" Sullivan
	see Sullivan, Imogene "Jean," 158–159
Bledsoe, John, 159
Bledsoe, Kori, 159
Bledsoe, Ricky, 159
Bledsoe, Wanda
	see Welch, Wanda Bledsoe, 159
Bledsoe, Willis, 159
Blue Grass Century Farm, 98
Blue Grass Farm, 59–61
Boiling Spring Academy and Mounds, 1
boll weevil, 136
Bond brothers, 180
Bond Farm, 97–98
Bond, Allen, 60, 98
Bond, Carol, 60
Bond, Carol Allen
	see Allen, Carol, 98
Bond, Charles, 60, 98
Bond, Cicero, 115
Bond, Cicero Columbus "CC," 60
Bond, Connor, 98
Bond, Cora Steele
	see Steele, Cora, 84, 115
Bond, Dan, 60, 98, 140
Bond, Dorothy, 125
Bond, Elise Core
	see Core, Elise, 124
Bond, Elizabeth J. "Bettie"
	see Anderson, Bettie, 139–140
Bond, Eva, 84
Bond, Gladys
	see Wilson, Gladys Bond, 123–125
Bond, James, 98
Bond, James W., Jr., 97–98
Bond, James William, 60
Bond, Jane
	see Giles, Jane Bond, 84
Bond, John, 83
Bond, John B., 59, 123–124
Bond, Leo Ratcliff Grigsby
	see Grigsby, Leo Ratcliff, 97–98
Bond, Leonard, 123–124
Bond, Loreen, 84–85, 115
Bond, Lucile, 123–125
Bond, Margaret Vantrease, 117
Bond, Mary, 84–85
Bond, Mary Elizabeth, 115
Bond, Phyllis, 140
Bond, Robert, 60
Bond, Tom, 83
Bond, William Cicero, 84–85, 115
Bond, William Franklin, "Banny," 84
Bond, William Howard, 84–85
Bostick, Bethenia
	see Patton, Bethenia Bostick, 43
Bostick, John
Revolutionary War veteran, 43
Bostick, John and Mary Jarvis, 43
Bowersock, Bill, 11
Bowersock, Lola Reed Glenn
	see Glenn, Lola Reed, 11–12
Bowman, Virgiinia, 24, 78, 81, 189
Bowman, Virginia, 24, 81, 84
Boyd family, 35
Boyd Mill Pike, 35–36
Boyd, Tennie, 36
Boyd, William Anderson, 36
Boyd, William Irby and Margaret Anderson, 35
Boyd's Distillery, 36
Boyd's Mill, 36
Branham and Hughes Military Academy, 99
Brentwood Country Club and Golf Course, 180
Broken Bow Construction Co., 150
Brown, Col. Hugh, 129
Brown, Daniel, 15
Brown, Elizabeth
	see Lee, Elizabeth Brown, 15
Brown, Mary Polk
	see Polk, Mary, 15
Brown, Virginia "Jennie"
	see Pointer, Virginia Brown, 129
Broyles, Martha
	see Lee, Martha Broyles, 18
Bryan, Elizabeth, see Bond, Elizabeth Bryan, 59

Bud's Longview Farm, 139–141
Bur (slave), 11
Burns, Angela, 188
Burns, Carolyn, 188
Burns, Flora Church
    see Church, Flora, 51
Burns, George, 51
Burns, James, 51
Burns, Martha, 188
Burns, Sara, 188
Burns, Sophronia Ann
    see Peach, Sophronia Ann Burns, 51
Burns, Thomas, 188
Burns, Virginia
    see Poyner, Virginia Burns, 51
Burns, William, 51
Burwood, 106, 120, 172
Burwood and Beyond (title), 127
Byrd, Betty Plant
    see Plant, Betty, 172–173
Byrd, Debra Lynn, 173
Byrd, Hannah, 173
Byrd, Phillip W. Byrd, Sr., 173
Byrd, Phillip, Jr., 173
Byrd, Rachel Kathryn "Katie" Johnson
    see Johnson, Rachel Kathryn "Katie," 172
Byrd, Thomas Jefferson, Jr., 172–173
Byrd, Thomas Jefferson, Sr., 172–173

C
Caldwell, Dr. St. Clair, 129
Caldwell, Martha
    see Pointer, Martha Caldwell, 129
Camp Trousdale, 69
Campbell family, 88
Campbell, Capt. McCoy Clemson, 130
Campbell, Henry, 130
Campbell, John, 51
Campbell, Mary Louise
    see Polk, Mary Louise Campbell, 130
Campbell, Mattie "Patsy"
    see Pointer, Mattie "Patsy" Campbell, 130
Caney Fork, 190
Cannon Century Farm, 65–67
Cannon Farm, 65–67
Cannon, Ed, 66
Cannon, Edgar Brown, 66
Cannon, Marguerite, 66
Cannon, Newton
    Governor of Tennessee, 65–66
Cannon, Rachel Adeline
    see Maney, Rachel Adeline Cannon, 66
Cannon, Susan Agatha Perkins
    see Perkins, Susan Agatha, 65
Cannon, Virginia Brown McEwen
    see McEwen, Virginia Brown, 66

Cannon, William Perkins, 65–66
    see also Perkins, Susan Agatha, 65–66
Carson, Byron Carvell, 133
Carson, Lisa, 133
Carson, Rondell Stacey, 133
Carson, Virginia
    see Jefferson, Virginia Carson, 81
Carson, Zelda Pewitt
    see Pewitt, Zelda, 132–133
Carter House, Franklin, 182
Carter, Alice Watson
    see Watson, Alice, 187
Carter, Fountain Branch, 182
Carter, Fred, 187
Carter, Jeff, 187
Carter, John C., 182
Carter, Lottie
    see Ashworth, Lottie Carter, 187
Carter's Acres Farm, 187
Cartwright Farm, 188
Cathey, Sallie P.
    see Smith, Sallie P. Cathey, 102
Cedar Creek Farm, 117–119
Cedar Lane Farm I, 189
Cedar Lane Farm II, 189
Cemeteries
    Cannon Century Farm, 65–67
    Cross Keys, 117
    Eastview, 126
    Green Chapel, 133
    Lester, 63
    Moses Steele, 11, 74
    Parker-Fowlkes Cemetery, 159
    Smith Family Cemetery, 102
Center for Historic Preservation (CHP)
    Middle Tennessee State University (MTSU), 3, 4
Central State Hospital, Nashville, 70
Century Farms
    County Fairs, 3
    Signs, 3–4
Century Farms Program
    application and proof, 3
Channell, Hugh, 78
Channell, Rebecca
    see Gentry, Rebecca Channell, 78
Chapman, Nan
    see Rodgers, Nan Chapman, 182
Charles Gentry Farm, 112–114
Cherokee, 2
Chester, Lynn
    see Wilson, Lynn Chester, 124–125
Chickasaw, 2
Childress, Sarah
    see Polk, Sarah Childress, 11
Choctaw, 2

Chreisman, Martha E.
    see Hatcher, Martha E. Chreisman, 69
Church, Flora
    see Burns, Flora Church, 51
Civil War, 15, 16, 24, 36, 39, 40, 46, 47, 59, 60, 66, 69, 74, 77, 81, 84, 87–89, 95, 154, 184, 186
    body servant, 187
    prisoner of war, 129
Clark, Helen Ann, 166
Clark, Wiliam P., 166
Cleburne Farm, 88
Cleburne Jersey Farm, 130
Coast Artillery, 142
Coleman, Elizabeth Luviney
    see McCanless, Elizabeth Luviney Coleman, 95
College Grove, 20, 99
College Grove School, 153
Confederate Army, 17, 129, 149
Connecticut, Salisbury, 15
Connell, Kerry, 62–63
Connell, Mary Margaret Sanford
    see Sanford, Mary Margaret, 63
Connell, Sharon, 62–63
Connell, Terry L., 63
Core, Elise
    see Bond, Elise Core, 124
County Line Farm, 91–94
Craig, Nancy Barker
    see Barker, Nancy, 121–122
Creek, 2
Creek War, 66
Creekside, 185
Crestview Farm
    aerial view, 174
Creswell, Mildred
    see McCall, Mildred Creswell, 74
Cross Keys community, 117
Crowson, Andrew Jackson, 120
Crowson, Lou Ella, 120
Crowson, Prentice, 120
Crowson, Telitha Barker
    see Barker, Talitha, 120
Crowson, Willie, 120
Crystal Valley Century Farm, 112
Crystal Valley Farm, 95–96
Cumberland University, 17
Cumberland University Law School, 70
Cumberland Valley, 7
Cummins, Elizabeth Lewis
    see Herbert, Elizabeth Lewis Cummins, 57

D
Dairy industry, 136
Dairy industry in Tennessee, 160
Dairy, Grade A, 60, 188
Dairy, Grade B, 127
David Lipscomb College, 92
Deal, Lenar Florence
    see Sullivan, Lenar Florence Deal, 105
DeGrafenried family, 77
Demonbreun, Willie
    see Patton, Willia Demonbreun, 42–43
Depression, the Great, 135, 183, 184
Devlin Farms, 44
Dickson, Dickson County, 109
Dodson, Mary Susan
    see Hatcher, Mary Susan Dodson, 69
Dotson, Ann, 188
Dotson, Benjamin, 188
Dotson, Carolyn
    see Burns, Carolyn Dotson, 188
Dotson, Sylvester, 188
Dotson, Walter, 188
Douglas, Mattie Reeves, 42
Douglass Chapel, 80
Douglass-Reams House, 80
Douglass, Frances, 80–81
Douglass, Thomas Logan, 80
Dripping Springs, 186
Duncan, Polly Barker
    see Barker, Polly, 120
Duplex, community, 14, 19
Duplex, horse, 17, 19

E
Edmondson, Frances
    see Linton, Frances Edmondson, 92
Edwards Gove (church), 153
Egbert, Camille Pearre, 168–169
Egbert, Paul Martin, 168–170
Egbert, Rose Elaine, 168–170
Egbert, Virginia Pearre
Pearre, Virginia, 168–170
Ernst, Christian, 18
Ernst, Eli, 18
Ernst, Laura Lee
    see also Lee, Laura, 18
Ernst, Martha, 18
Europeans, 2
Ewell, Richard, 88

F
Fain, Margaret
    see Moran, Margaret Fain, 183
Fairview, 109–111, 149, 158–159, 190
Fairview City Manager, 110
Fairview Elementary School, 132
Farm Bureau, 84, 143, 145
Farm hand, unnamed, 33
Farm machinery, 90–92
Farm Service Agency, 147

Farnsworth, S.E., 66
Fernvale, Tennessee, 91
Fewkes Site, 1
FFA (Future Farmers of America), 103
First District, 109
Fisher, Elvana Anderson
    see Anderson, Elvana, 166
Fisher, G.B., 166
Fisher, Nancy Jane Sullivan
    see Sullivan, Nancy Jane, 166
Fisher, Steve, 73–74
Fisher, Susan McCall
    see also McCall, Susan, 73–75
Fisher, W.J., 166
Fitts, Stephen and Alicia, 131
Flat Creek, 10, 11, 74, 152–153
Flat Creek Community Club, 153
Flowers, Parmelia Caldonia "Callie
    see Bledsoe, Parmelia Caldonia Flowers "Callie," 158–159
Fly, Will, 161
Forrest, Nathan Bedford, 66
Fourth Tennessee Cavalry, 84
Fox's Sawmill, 170
Franklin, 185
Franklin High School, 175
Franklin Preservation Partners, 185
Freed men and women, 154
Future Farmers of America, 153

G
Garrison community, 51, 181
Gentry Farm, 76–79
Gentry Farm Fall Pumpkin event, 76, 79
Gentry, Allen, 76, 78, 79
Gentry, Charles, 112–113
Gentry, Cindy White
    see White, Cindy, 76, 78–79
Gentry, Dorothy, 113
Gentry, Elliott Brown, 113
Gentry, Hope, 79
Gentry, James, Jr., 78
Gentry, Jase, 79
Gentry, Jean, 113
Gentry, Jimmy, 78
Gentry, Katherine, 113
Gentry, Margaret Lampley
    see also Lampley, Margaret, 113
Gentry, Martha Virginia, 113
Gentry, Mary McCanless
    see also McCanless, Mary, 112
Gentry, Mary Morgan
    see Hatcher, Mary Morgan Gentry, 68, 71, 79
Gentry, Mary Ruth, 113
Gentry, Minnie Eugenia Green
    see also Green, Minnie Eugenia, 113
Gentry, Rebecca Channell
    see also Channell, Rebecca, 78
Gentry, Scott, 78
Gentry, Wayne, 113
German Farm, 24–28
German, Cynthia
    see Williams, Cynthia German, 25
German, Daniel, 24
German, Emaline McEwen
    see McEwen, Emaline, 24–25
German, Fanny Puckett, 24
German, Sarah, 24
German, Zacheus, 24–25
Gibbs, Alphonso, 178
Gibbs, Elizabeth
    see Moss, Elizabeth Gibbs, 184
Gibbs, Mrs. Kennedy (Moss) Gibbs
    see also Moss, Elizabeth, 184
Giddens, Francis, 178
Giddens, Sarah
    see Moss, Sarah Giddens, 184
Giles, Corey, 84, 115–116
Giles, David, 84, 115–116
Giles, Gary, 84
Giles, Jane
    see Smithson, Jane Giles, 74
Giles, Jane Bond, 115–116
    see also Bond, Jane, 84
Giles, Jill, 115–116
Giles, Luke, 115–116
Gillespie Place
    see Glenn Acres, 10
Gillespie, David, 11, 74
Gillespie, David and Anna, 11
Gillespie, George, 11
Gillespie, Isaac, 11, 74
Gillespie, Lydia
    see Knox, Lydia Gillespie, 11
Gillespie, Mary Ann McGuire, 11
Gillespie, Mary Armenia
    see Williams, Mary Armenia Gillespie, 25
Gillespie, Pauline Stephens
    see Stephens, Pauline, 11
Gillespie, Samuel, 11
Gillespie, Thomas, 11
Gillespie, Thomas and Naomi, 10–11
Gillespie, Thomas, Jr., 74
Gillespie, Wallace, 11
Gillespie, William Henry "Bill," 11
Givens Family
    Grady B. and Stacey L. Givens Trust, 150
Givens, Grady (5th gen), 150–151
Givens, Grady B., 150–151
Givens, Grady McClelland, 150
Givens, Johnnie Russell, "JR," 190

Givens, Kern, 150–151
Givens, Lucinda Sullivan
    see Sullivan, Lucinda, 190
Givens, Ora Sullivan
    see Sullivan, Ora, 150
Givens, Sidney Pollard, 190
Givens, Stacey, 143
Givens, Stacey Williams
    see Williams, Stacey, 150–151
Givens, Wilma Dean Potchad
    see Potchad, Wilma Dean, 150
Glass Mounds, 1
Glass, Agnes Hunter
    see also Hunter, Agnes, 77
Glass, Corinne
    see Gordon, Corinne Glass, 78
Glass, Samuel F., 77
Glass, Samuel F., Jr., 77
Glass, Sarah Malone, 77
Glenn Acres
    see Gillespie Place, 10
Glenn Acres Farm, 7, 10–13
Glenn, Calvin, 12–13
Glenn, Jackie, 12
Glenn, Janice, 12
Glenn, Lola Reed
    see also Bowersock, Lola Reed Glenn, 11–12
    see also, Reed, Lola, 11
Glenn, Martin Dodson "Jackie," 11
Glenn, Sandra Thompson
    see Thompson, Sandra, 12–13
Glenn, Susan, 12
Glenn, Tony, 12
Glenn, Tucker, 13
Goolsby, Harper, 153
Goolsby, Lindsey, 153
goose house, 190
Gordon, Corinne Glass
    see also Glass, Corinne, 78
Gordon, Edward Allen, 78
Gosey, Minnie Pearl
    see Dotson, Minnie Pearl Gosey, 188
Gould, James, 16
Granville County, North Carolina, 126
Grassland community, 29
Grassland Reserve Program, 150
Green Brothers Dairy, 41
Green, Allen, 41
Green, Ben and Janice, 41
Green, Bettie Battle
    see also Battle, Bettie, 41
Green, Bill, 41
Green, Edward, 39
Green, Edward and Mary, 39, 41
Green, Ethel McMahon
    see McMahon, Ethel, 41
Green, Jennifer, 41
Green, John Edward (John Ed), 41
Green, Lundy, 39
Green, Lundy and Maude, 39
Green, Maude York
    see York, Maude, 39, 41
Green, Minnie Eugenia
    see Gentry, Minnie Eugenia Green, 113
Green, Sherwood, Jr., 39
Grigsby family, 97
Grigsby, Annie Lou Barker
    see Barker, Annie Lou, 120, 127
Grigsby, Charles, 98
Grigsby, Charles F., 98
Grigsby, Cleo, 180
Grigsby, Ella Frances, 98
Grigsby, Ethel, 98
Grigsby, Harry, 98
Grigsby, Jack, 127
Grigsby, Judith
    see Hayes, Judith Grigsby, 127
Grigsby, Judy, 127
Grigsby, Katherine, 98
Grigsby, L.B., 127
Grigsby, Leo Ratcliff
    see Bond, Leo Ratcliff Grigsby, 97–98
Grigsby, Marion, 98
Grigsby, Scales, 98
Guernsey cattle, 88

H
Hardcastle, Susan Jefferson
    see also Jefferson, Susan, 82
Hargrove, A.F., 117
Hargrove, Billie Jean
    see Lillard, Billie Jean Hargrove, 118
Hargrove, Effie
Scales, Effie Hargrove, 184
Hargrove, Mary Louise Trice
    see Trice, Mary Louise, 117–118
Harper, Allean
    see Wallace, Allean Harper, 152–153
Harpeth Conservancy awards
Agriculture Award
    Harpeth River Steward Award, 175
Harpeth River, 1
Harpeth River Steward Award, 175
Harpeth River Steward Award and Agriculture Award
    Harpeth Conservancy, 175
Hatcher Family Dairy, 68–72
Hatcher, Abram Wooldridge, 69
Hatcher, Caledonia Pillow
    see also Pillow, Caledonia, 69
Hatcher, Cannon, 68, 71, 72

Hatcher, Charles, 68, 71, 72
Hatcher, Charlie, 147
Hatcher, Dr. Charlie, 68, 71
Hatcher, Dr. Jennifer, 71
Hatcher, Elder, 148
Hatcher, Era Emma "Sadie"
    see also Kinnard, Era Emma "Sadie," 147–148
Hatcher, Eula Neely
    see Neely, Eula, 69–70
Hatcher, George, 69
Hatcher, George Abram (Abe), 70–71
Hatcher, Jacqueline
    see Price, Jacqueline, 70–71
Hatcher, Jasper G. Memorial Highway
    see Jasper G. Hatcher Memorial Highway, 148
Hatcher, Jasper, Jr., 148
Hatcher, Jasper, Sr., 147
Hatcher, Jessie Mary, 147
Hatcher, Jimmie, 147
Hatcher, Laeunia, 147
Hatcher, Lawrence Buford, Sr., 148
Hatcher, Lucy
    see Miller, Lucy Hatcher, 69
Hatcher, Mariah, 147
Hatcher, Martha E. Chreisman
    see also Chreisman, Martha E., 69
Hatcher, Martha Jane, 72
Hatcher, Marvin, Sr., 147–148
Hatcher, Mary Morgan Gentry
    see also Gentry, Mary Morgan, 68, 71, 79
Hatcher, Mary Susan Dodson
    see also Dodson, Mary Susan, 69
Hatcher, Meredith, 146–147
Hatcher, Ned, 146
Hatcher, Octavius Claiborne, "O.C.," 69
Hatcher, Rosa Moss
    see also Moss, Rosa, 146
Hatcher, Sharon, 68, 71, 72
Hatcher, Spotswood Henry, 69
Hatcher, Thelma Thressa, 148
Hatcher, Thomas Logwood, 69
Hatcher, Vinnie Starnes
    see Starnes, Vinnie, 146
Hatcher, William and Lucy, 69
Hatcher, Winnie, 147
Hawkins, Jacqueline "Jackie" Reid
    See, Ogilvie, Jacqueline Reid Hawkins, 100
Hayboer, Sandra Barker
    see Barker, Sandra, 121–122
Hayes, Judith Grigsby
    see Grigsby, Judith, 127–128
Hazelwood, Annie
    see Smith, Annie Hazelwood, 102–103
Hen House, The Old, 13

Henderson, Samuel, 74
Herbert, Bob, 55–56
Herbert, David Cummins, 57
Herbert, Elizabeth Lewis Cummins
    see Cummins, Elizabeth Lewis, 57
Herbert, George O., Sr. ("Buck"), 56–57
Herbert, George Washington Sneed, 57
Herbert, Gladys, 56
Herbert, James Harvey, 57
Herbert, John Overton, 57
Herbert, Mary Selina, 57
Herbert, R.N., II ("Roose"), 56
Herbert, Richard, 55
Herbert, Robert N., 55–57
Herbert, Robert N., Jr., 57
Herbert, Robert Nathaniel, 57
Herbert, Shelby, 13
Herbert, Thomas, 55
Herbert, Thomas L., 57
Herbert, Tom, 55–56, 58
Herbert, Woody, 55
Heritage Foundation History Classroom, 79
Highway Commission, 143
Hill, Kristi Roberta
    see Jefferson, Kristi Roberta Hill, 82
Hillsboro / Leiper's Fork, 107
Hillsboro High School, 143
Hillsdale Farm, 181
Holstein Friesian cattle, 88
Holt, Clay, 103
Holt, Jeff, 102–103
Holt, Jennifer, 103
Holt, Stacey, 103
Home Demonstration Club, 103, 136, 143, 145, 150, 186
Hughes family, 168
Hughes, Betty Dean
    see, Lampley, Betty Dean Hughes, 110–111
Hull, Frank, 107
Hunt-Beasley Farm, 106–108
Hunt, Ella Beasley
    see also Beasley, Ella, 106–107
Hunt, J. Buchanan "Buck," 106–107
Hunter, Agnes
    see Glass, Agnes Hunter, 77
Hunter, Chris, 38
Hunter, Christopher, 36
Hunter, Dustin Short, 36
Hunter, Farley, 36
Hunter, Floyd, 36
Hunter, Harriet
    see McCullough, Harriet Hunter, 179
Hunter, Henry, 179
Hunter, Jeffrey, 36
Hunter, Nicholas Alexander, 36
Hunter, Robbie, 179

Hunter, Vickie, 36
Hunter, Virginia Short, 36

I

Iron Furnace
    Lee and Gould Furnace, 16

J

J.R. Givens, LLC, 190
Jasper G. Hatcher Memorial Highway
    see Hatcher, Jasper G. Memorial Highway, 148
Jaworski, Brandon, 13
Jefferson Farms, 80–82
Jefferson, Adam, 82
Jefferson, Caswell M., 81
Jefferson, Caswell M., III, 82
Jefferson, Caswell M., Jr., 82
Jefferson, Jonathan Willard, Sr., 81
Jefferson, Joyce M. Waggoner
    see Waggoner, Joyce M., 82
Jefferson, Kristi Roberta Hill
    see also Hill, Kristi Roberta, 82
Jefferson, Robert Reams, 81–82
Jefferson, Sallie Reams
    see Reams, Sallie, 81
Jefferson, Susan
    see Hardcastle, Susan Jefferson, 82
Jefferson, Virginia Carson
    see also Carson, Virginia, 81
Jenkins, Agnes
    see Nichols, Agnes Jenkins, 161
Jenkins, Emmitt, 161
Jim (slave), 11
Jingo Post Office, 142
John Deere, 107
Johnson, Ardeen McCanless
    see also McCanless, Ardeen, 96
Johnson, Cliff, 111
Johnson, Jacqueline May, 82
Johnson, James Knox Polk, 96
Johnson, Jefferson, 82
Johnson, Jeremy, 82
Johnson, John, 96
Johnson, Julianna, 82
Johnson, Rachel Kathryn "Katie"
    see Byrd, Rachel Kathryn "Katie" Johnson, 172
Johnson, Stephanie Reams
    see Reams, Stephanie, 82
Johnson, Stuart, 111
Johnson, Vicki, 111
Johnson, William Hazlewood, 96
Johnsongrass Farm
    see Lampley Farm, 109–111
Jones, Alice
    see Sparkman, Alice Jones, 107–108
Jones, Joseph Ridley, 107
Jones, Joseph William "Jay," 107
Jones, Larry, 121
Jones, Sallie
    see Luster, Sallie Jones, 155
Jones, Willie Beasley
    see also Beasley, Willie, 107
Jordan, Annie Page
    see Page, Annie, 144
Jordan, Archer Lee, "A.L.," 144
Jordan, Aubrey, 144
Jordan, Bruce, 144
Jordan, Dr. Lee, 145
Jordan, Emma Ida
    see Wilson, Emma Ida, 144
Jordan, Garner, 144
Jordan, Nannie Sue, 144
Jordan, Thomas Lee, 144
Jordan, Walter, 144
Jordan, Walter C, 144–145
Jordan, Walter Lee, 144–145
Jordan, William, 144

K

Kester, Emily, 13
Kester, Mason, 13
Kester, Stella, 13
Kienzle, Valerie Battle
    see also Battle, Valerie, 23
King, John W., 39
King's Chapel development, 43
Kinnard, Era Emma "Sadie"
    see Hatcher, Era Emma "Sadie," 147–148
Kinnie, John, 183
Kinnie, Paul, 183
Kirby, Alice
    see Meacham, Alice Kirby, 181
Knox, Jane
    see Polk, Jane Knox, 11
Knox, John, 11
Knox, Lydia Gillespie
    see also Gillespie, Lydia, 11
Korean War, 84

L

Lampley Farm, 109–111
    see also Johnsongrass Farm, 109–111
Lampley, Anthony, 111
Lampley, Betty Dean Hughes
    see also Hughes, Betty Dean, 110–111
Lampley, Earl Demarquis, Jr., 110
Lampley, Earl Demarquis, Sr., 110–111
Lampley, Margaret
    see Gentry, Margaret Lampley, 113
Lampley, Mark, 111

Lampley, Molly, 110
Lampley, Uzebe, 110
Lampley, Vickie, 111
Land Grant, Revolutionary, 15
Land Grants, 6, 7, 11, 15, 59, 65, 167
Land Trust for Tennessee, 33, 34, 73, 75, 174, 189
Lavender, Delilah
    see Shaw, Delilah Lavender, 126
Lee, Charles Alfred, 16–18
Lee, Elizabeth Brown
    see also Brown, Elizabeth, 15
Lee, Florence Amanda, 16–17
Lee, John
    see Lee, John Napier, 14, 17, 18
Lee, John McCutcheon "Jack," 18
Lee, John Napier III, 18
Lee, John Wills Napier, 16–17
Lee, John Wills Napier Lee, Jr., 17
Lee, Laura
    see Ernst, Laura Lee, 18
Lee, Martha Broyles
see also Broyles, Martha, 18
Lee, Mason, 18
Lee, Miller, 18
Lee, Mona Robertson
    see also, Robertson, Mona, 17–18
Lee, Nancy Anderson, 18
Lee, Sam L., 18
Lee, Samuel, 15
Lee, Samuel Brown, 15–16
Lee, Samuel Brown, Jr., 16–17
Leiper's Fork, 179
Leiper's Fork Distillery, 179
Lieutenant Governor's Staff, Honorary Member, 148
    Lillard, Billie Jean Hargrove
        see also Hargrove, Billie Jean, 118
Lillard, Diane
    see Marlin, Diane Lillard, 118–119
Lillard, James Marvin, 118
Linton, Frances Edmondson
    see also Edmondson, Frances, 92
Linton, Hooper, 91–92, 94
Linton, Ida McPherson
    see McPherson, Ida, 91
Linton, Katie Anderson
    see Anderson, Katie, 91
Linton, Lloyd, 92
Linton, Mary
    see Little, Mary Linton, 91, 93–94
Linton, Silas, 91, 94
Linton, Tennessee, 91
Linton, Will, 91
Little family
    Ben, Mary, Andrew, Elizabeth, Brittain, Shelley, Tess, and Hudson, 92

Little, Andrew, 93
Little, Barbara Jean
    see McCanless, Barbara Jean Little, 96
Little, Ben, 91, 93
Little, Brittain, 93
Little, Mary Linton
    see also Linton, Mary, 91, 93–94
Lloyd Linton Bridge, 92
Lloyd, Ava Claire, 118
Lloyd, Camden, 118
Lloyd, Jennifer Marlin
    see also Marlin, Jennifer, 118
Lloyd, Trey, 118
Locust Guard, 29–32
Long View Farm, 129–131
Loreen and Mary Bond Farm, 115–116
Love, Frances
    formerly Frances Douglass, 80–81
Lovell, Susan Jane
    see McCanless, Susan Jane Lovell, 95
Luster family, Caden, 154
Luster family, Desarae, 157
Luster family, Jordan, 156
Luster family, Kynnedi, 156
Luster Farm, 154–157
Luster Farm petition, 155–156
Luster-Smith, Erica, 155–157
Luster-Vinson, Tasha, 155–156
Luster, Amber, 155–157
Luster, Anna, 155
Luster, Anthony W., 155–157
Luster, Betsey, 154
Luster, Cynthia, 155
Luster, Grant, Jr., 154–155
Luster, Grant, Sr., 155
Luster, James, 155
Luster, Jennifer, 155–157
Luster, Kendra, 155–157
Luster, Mattie Jane Smithson
    see Smithson, Mattie Jane, 155
Luster, Nellie, 155–157
Luster, Nelson, 154
Luster, Nelson II, 155
Luster, Sallie Jones
    see also Jones, Sallie, 155
Luster, William, 155

M
Magistrate, 1st District, 150
Mammy Ann, 16
Maney, Rachel Adeline Cannon
    see also Cannon, Rachel Adeline, 66
Maple Crest Farm, 99–101
Maple Lawn Farm, 125
Maplewood, 14–19
Maplewood Farm, 7

Marlin, Amanda
    see Pilkinton, Amanda Marlin, 118
Marlin, Diane Lillard
    see also Lillard, Diane, 118–119
Marlin, Gene, 118
Marlin, Jennifer
    see Lloyd, Jennifer Marlin, 118
Mary and Loreen Bond Farm, 84, 117
Maury County Co-Op, 130
Mayflower, 30
Mayor of Fairview, 111
McCall, Alice Smithson
    see Smithson, Alice, 74
McCall, Andrew Lycurgus, 74
McCall, Col. Gerald, 74
McCall, Herbert Lycurgus, 74
McCall, Mildred Creswell
    see also Creswell, Mildred, 74
McCall, Susan
    see Fisher, Susan McCall, 73–75
McCanless family, 112
McCanless, Ardeen
    see Johnson, Ardeen McCanless, 96
McCanless, D. Brown, 96
McCanless, Elizabeth Luviney Coleman
    see also Coleman, Elizabeth Luviney, 95
McCanless, James Caldwell, Sr., 96
McCanless, James Thomas Carroll, 95–96
McCanless, James, Jr., 96
McCanless, Jonathan Lee, 96
McCanless, Mary
    see also Gentry, Mary McCanless, 112
McCanless, Nina June, 96
McCanless, Robin Carol
    see Thomas, Robin Carol McCanless, 96
McClaran, Nannie
    see Page, Nannie McClaran, 144
McClellan, Sarah Frances "Fannie"
    see Ring, Sarah Frances McClellan, 30
McCord, Annie Lou Reed
    see Reed, Annie Lou, 189
McCord, Dorothy
    see Ryan, Dorothy McCord, 189
McCord, Elizabeth
    see Crunk, Elizabeth McCord, 189
McCord, Walker L., 189
McEwen, Emaline
    see German, Emaline McEwen, 24–25
McEwen, Florence
    see Rosser, Florence McEwen, 185
McEwen, John B. and Cynthia, 185
McEwen, Virginia Brown
    see Cannon, Virginia Brown McEwen, 66
McFarlin, Benajah Hill, 161
McFarlin, Lena Senethius
    see Nichols, Lena Senethius McFarlin, 161

McGavock family, 77
McGavock, Lysander and Elizabeth, 180
McMahon, Ethel
    see Green, Ethel McMahon, 41
McMurray, Mary
    see Osburn, Mary McMurray, 175
McPherson, Ida
    see Linton, Ida McPherson, 91
McWherter, Ned Ray
Speaker of the House, 148
Meacham family, 181
Meacham, Alice Kirby
    see also Kirby, Alice, 181
Meacham, Florence
    see Pewitt, Florence Meacham, 186
Meacham, Harold, 181
Meacham, John and Elizabeth, 186
Meacham, M.A. and Alice Kirby Meacham, 181
Meacham, Margherita, 186
Meacham, Milton and Pina Jane, 186
Meadow Gold Dairy, 113
Methodist Circuit Rider, 51
Methodist Ministers, 80
Mexican War, 69
Middle Tennessee State University, 1, 3, 98, 113, 142
Middle Tennessee State University (MTSU) Center for Historic Preservation (CHP), 3, 4
Midway Farm, 180
Miller, Lucy Hatcher
    see also Hatcher, Lucy, 69
Miller, Megan
    see Lee, Megan Miller, 18
Miller, Rucker, 69
Mississippian Period, 1
Mitchum, Alice Ann, 130, 131
Mitchum, Mary Polk
    see also Polk, Mary, 130
Mitchum, Millard "Bud," 130
Mitchum, Millard, Sr., 130
Mockingbird Hill, 184
Montrose Farm, 180
Moore, Susan
    see Steele, Susan Moore, 83
Moran family, 183
Moran, Ann, 183
Moran, James, 183
Moran, Margaret Fain
    see also Fain, Margaret, 183
Moran, Samuel Houston, 183
Mosley, Era Jane
    see Sanford, Era Jane Mosley, 62
Moss Side Farm, 178
Moss, Elizabeth
    see Gibbs, Mrs. Kennedy (Moss) Gibbs, 184
Moss, Elizabeth Gibbs

see also Gibbs, Elizabeth, 184
Moss, Evie, 178
Moss, Rosa
 see Hatcher, Rosa Moss, 146
Moss, Sarah Giddens
 see also Giddens, Sarah, 184
Moss, William, 184
Motheral, Anness, 29
Motheral, Emma Tennessee
 see Ring, Emma Tennessee Motheral, 29–30
Motheral, Jane Currie, 29
Motheral, John, 29
Motheral, Joseph, 29
Mounds, 1
MTSU
 see also Middle Tennessee State University, 78, 79
Murfree, Col. Hardy, 65
Murfree, William, 65
Murfreesboro, 65

N

Napier, John, 15
Napier, Susan
Lee, Susan Napier, 15–17
Nashville Central High School, 56
Nashville Electric Service, 173
Nashville-Hillsboro Turnpike (Hwy 431), 36
Natchez Trace, 1, 183
National Bank of Franklin, 183
National Register of Historic Places, 1, 14, 21, 41, 77, 80, 126, 179, 185
National Resources Conservation Service, 175
Native Americans, 6–7
Native grasses
 Blue Stem
 Little Blue Stem
 Switch Grass, 104
Neely, Eula
 see Hatcher, Eula Neely, 69–70
Nelson Creek, 43
New Hope School, 143
Nichols Family, four generations, 163
Nichols Farm, 160–164
Nichols Jersey Farm, 160
Nichols, Agnes Jenkins
 see also Jenkins, Agnes, 161
Nichols, Berry D., 161
Nichols, Brandon, 161–162
Nichols, Brittany, 161, 164
Nichols, Douglas, 161
Nichols, Herbert, 161–162
Nichols, Kyle, 161
Nichols, Leigh Ann, 161
Nichols, Lena Senethius McFarlin
 see McFarlin, Lena Senethius, 161
Nichols, Mark, 161–164
Nichols, Mary Jane, 161
Nichols, Rebecca, 161
Nichols, Robert, 161
Nichols, Sandra, 161–162
Nichols, Sue Mildred, 161
Nichols' granddaughter, Kendall, 164
Nichols' granddaughter, Myla, 164
Nichols' grandson, Emmitt, 164
Nichols' grandson, Landon, 164
Nix, Rose Ann
see Pearre, Rose Ann Nix, 169
Nolensville, 39, 95
Norman, Robbie Barker
 see also Barker, Robbie, 120
North Carolina
Rowan County, 10
North Carolina Land Act of 1783, 6
North, Elizabeth
 see Reams, Elizabeth North, 81
Nutt, Donna
 see Bledsoe, Donna Nutt, 159

O

Oaklands Mansion, 66
Ogilvie Farm
 see Beech Hill, 20
Ogilvie, Anna Rucker
 see also, Rucker, Anna, 21
Ogilvie, Annie Lou, 100–101
Ogilvie, Bettye Maxwell, 21
Ogilvie, Cynthia, 20
Ogilvie, Elizabeth
 see Battle, Elizabeth Ogilvie, 22–23
Ogilvie, Elizabeth Maxwell, 21
Ogilvie, Jacqueline Reid Hawkins
 see also Hawkins, Jacqueline "Jackie" Reid, 100
Ogilvie, James D., 21
Ogilvie, James Smith, 20–21
Ogilvie, James Smith II, 21
Ogilvie, Kathleen Smith
 see Smith, Kathleen, 99–100
Ogilvie, Lynn
 see Woodside, Lynn Ogilvie, 100
Ogilvie, Mary Helen, 100
Ogilvie, Rachel, 21
Ogilvie, Richard, 20
Ogilvie, Samuel, 21
Ogilvie, Samuel Jason, 21
Ogilvie, Samuel R., 22
Ogilvie, Samuel Rucker, 21
Ogilvie, Walter William, 99–101
Ogilvie, William "Bill" Harris, 100
Ogilvie, William and Mary Harris, 20

Ogilvie, William Harris and Annie Lou, 99, 101
Old Town, 1
Osburn, Leslie
    see Stallings, Leslie Osburn, 175
Osburn, Mary Louise
    see Stallings, Mary Louise Osburn, 174
Overbey, Era Frances
    see Walker, Era Frances Overbey, 143
Owen Hill Masonic Lodge, 153
Owen, Jane, 143
Ozburn Hessey, 34
Ozburn Hollow, 33–34
Ozburn, Alma, 33
Ozburn, Dessie, 33
Ozburn, Elaine, 33
Ozburn, Ellen, 33
Ozburn, F. Perry, Jr., 33
Ozburn, Jack, 33
Ozburn, John, Sr., 33
Ozburn, Leslie, 33
Ozburn, Mildred, 33
Ozburn, Robert and Jane Wylie, 33–34
Ozburn, Ruth, 33
Ozburn, Uncle Will, 33
Ozburn, Willie Dean, 33

P
Page High School, 103
Page Schools, 153
Page, Annie
    see Jordan, Annie Page, 144
Page, Nannie McClaran
    see also McClaran, Nannie, 144
Page, Thomas Harvey "T.H.," 144
Patron, Mattie Reeves, 42
Patton Road, 43, 45
Patton, Bethenia Bostick
    see also Bostick, Bethenia, 43
Patton, Carl, 42
Patton, Jason, 43
Patton, Mary Cleo Roberts
    see Roberts, Mary Cleo, 42
Patton, Mary Jane, 43
Patton, Mattie, 43
    see Stewart, Mattie Patton, 43–44
Patton, Minnie Aurelia, 42–44
Patton, Willia Demonbreun
    see also Demonbreun, Willie, 42–43
Patton, William, 42
Patton, William Douglas, 43
Patton, William Shannon, 42–43
Patton, William Shannon and Willie Demonbreun, 43
Patton, Willie, 42
Peabody College, 98, 142
Peaceful Valley, 152–153

Peach Hollow, 50
Peach, Annie Pearl, 51
Peach, Sophronia Ann Burns
    see Burns, Sophronia Ann, 51
Peach, W.S. and Sophronia, 50
Peach, William Strickland, 51
Pearre Creek Elementary School, 169
Pearre Farm, 168–171
Pearre, Buford, 168
Pearre, Edith, 168
Pearre, Eunice, 168–169
Pearre, Joe, Sr., 168
Pearre, Joseph Hunter, 169
Pearre, Milton, 168–169
Pearre, Paul, 168
Pearre, Paul H., 168–170
Pearre, Robert, 168–169
Pearre, Rose Ann Nix
    see Nix, Rose Ann, 169
Pearre, Virginia, 168
    see Egbert, Virginia Pearre, 168–170
Peck, Nancy
    see Walker, Nancy Peck, 142–143
Penn Hollow, 166
Penn Hollow Farm, 165–167
Penn, Robert, 166
People's School, 69
Perez, Blake, 13
Perez, Brady, 13
Perkins Branch, 167
Perkins family, 77
Perkins, Ebbin, 167
Perkins, Samuel and Nancy, 65
Perkins, Susan Agatha
    see Cannon, Susan Agatha Perkins, 65
    see Cannon, William Perkins, 65–66
Pettus, Nancy
    see Smithson, Nancy Pettus, 74
Pewitt Farm, 132–133
Pewitt, Bennie, 132–133
Pewitt, Earl, 132
Pewitt, Ephraim "E.W.," 132
Pewitt, Florence Meacham
    see Meacham, Florence, 186
Pewitt, Gentry, 132
Pewitt, Ida Dora
    see Sullivan, Ida Dora Pewitt, 104
Pewitt, John, 132
Pewitt, Kay
    see Vaughn, Kay Pewitt, 132–133
Pewitt, Lyn, 91, 110, 191
Pewitt, Lyn Sullivan, 110
Pewitt, Mary, 132
Pewitt, Zelda
    see Carson, Zelda Pewitt, 132–133
Pilkinton, Amanda Marlin

see also Marlin, Amanda, 118
Pilkinton, Caroline, 118
Pilkinton, Hattie, 118
Pilkinton, Jesse, 118
Pilkinton, Nan, 118
Pilkinton, Rob, 118
Pilkinton, Will, 118
Pillow, Caledonia
    see Hatcher, Caledonia Pillow, 69
Pinewood Ramblers, The, 132
Pioneer Century Farm, 14
Plant, Betty
    see Byrd, Betty Plant, 172–173
Pleasant View Farm, 76–79
Pointer family, 129
Pointer, Capt. Henry, 129
Pointer, Henry Strange, 129
Pointer, Martha Caldwell
    see also Caldwell, Martha, 129
Pointer, Mattie "Patsy" Campbell
    see also Campbell, Mattie "Patsy," 130
Pointer, Virginia Brown
    see also Brown, Virginia "Jennie," 129
Polk, Alice, 130
Polk, Allen, 130
Polk, Charles, 15
Polk, Dallis Shirling, 175
Polk, Eleanor, 175
Polk, Horace, 130
Polk, James, 15
Polk, James Knox, 11
Polk, Jane Knox
    see also Knox, Jane, 11
Polk, Josephine, 175
Polk, Mary
    see Brown, Mary Polk, 15
    see Mitchum, Mary Polk, 130
Polk, Mary Louise Campbell
    see also Campbell, Mary Louise, 130
Polk, Samuel, 11
Polk, Sarah Childress
see also Childress, Sarah, 11
Polk, Travis, 175
Pope, John, 126
Pope's Chapel, 126
Porter, Edward S., 3
Potchad, Wilma Dean
    see Givens, Wilma Dean Potchad, 150
Poyner Farm, 50–54
Poyner, Brenda, 51
Poyner, Chris, 51
Poyner, Claudine, 50–52
Poyner, Dewey, 53
Poyner, Douglas, 51–52
Poyner, Haley, 52
Poyner, Irene Yates, 51

Poyner, Jerry, 53
Poyner, Keith, 51
Poyner, Robert (Robbie), 51–52
Poyner, Tracy, 52
Poyner, Virginia Burns
    see also Burns, Virginia, 51
Poyner, W.T., 53
Poyner, Will, 52
Poyner, William Ewen, 51
Poyner, William Thomas, 51
Presley, Ella
    see Dotson, Ella Presley, 188
Price, Jacqueline
    see Hatcher, Jacqueline, 70–71
Progressive Farm Movement, 136
Puckett, Fanny
    see German, Fanny Puckett, 24

R
Reams-Jefferson and Jefferson Farms, 80–82
Reams, Elizabeth North
    see North, Elizabeth, 81
Reams, Robert, 80–81
Reams, Robert and Elizabeth, 82
Reams, Sallie
    see Jefferson, Sallie Reams, 81
Reams, Stephanie
    see Johnson, Stephanie Reams, 82
Rebuilding after the Civil War, 87–89
Reed, Annie Lou
    see McCord, Annie Lou Reed, 189
Reed, Elizabeth
    see Gillespie, Elizabeth Reed, 11
Reed, Lola
    see Glenn, Lola Reed, 11
Regent Homes, 41
Review-Appeal, 143
Revolutionary War, 66
Revolutionary War and Land Grants, 59
Revolutionary War veteran, 65, 178
Revolutionary War veterans, 74
Reynolds, James King, 184
Reynolds, Richard, 184
Reynolds, Susie Scales
    see Scales, Susie, 184
Ring, Andrew, 30–31
Ring, Beth, 30
[Ring], Daniel, 32
[Ring], Edie May, 32
Ring, Emma Mai, 30–31
Ring, Emma Tennessee Motheral
    see Motheral, Emma Tennessee, 29–30
Ring, Frank, 30
Ring, Henry Eleazar, 29–30
Ring, James E., 30
Ring, Jasper, 30

Ring, Ned, 30–31
Ring, Robert (Bob) and Charlene, 29–33
Ring, Sarah Frances McClellan
    see also McClellan, Sarah Frances "Fannie," 30
Rivers Meet Farm, 179
Road Commissioner, 140
Roberts, Mary Cleo
    see Patton, Mary Cleo Roberts, 42
Robertson, Mona
    see Lee, Mona Robertson, 17–18
Robotic milking system, 71
Rock Hill, 144
Rodgers, Nan Chapman
    see also Chapman, Nan, 182
Rodgers, Roche Carter, 182
Rodgerswood, 182
Rosser, Rev. William, 185
Rucker, Anna
    see Ogilvie, Anna Rucker, 21
Rutherford County, 11, 65
Rutherford, General Griffith, 11
Ryan, Dorothy McCord
see McCord, Dorothy, 189

S
Sally (slave), 11
Sanders, Harry, 179
Sanford, Albert, 62
Sanford, Archer Wood, 62
Sanford, Era Ann
    see White, Era Ann Sanford, 63
Sanford, Era Jane Mosley
    see also Mosley, Era Jane, 62
Sanford, Mary Ann Wood
    see Wood, Mary Ann, 62–63
Sanford, Mary Margaret
    see Connell, Mary Margaret Sanford, 63
Sanford, Minor, 62
Sanford, Robert, 62–63
Sanford, Sallie, 62
Sanford, Stephen, 62–63
Sanford, Viola Ward
    see Ward, Viola, 63
Saunders, Eddie, 31
Scales family, 97
Scales, Ella, 98
Scales, P.D. and Mary, 97
Scales, Reuben, 184
Scales, Susie
    see Reynolds, Susie Scales, 184
Settlers, 6–8
Shaw family, 120
Shaw, Delilah Lavender
    see also Lavender, Delilah, 126
Shaw, William A., 126

Shaw, William Augustus, 127
Shaw, William E., 126–127
Shawnee, 2
Sherwood Green Estates, 41
Sherwood Green Farm, 39–41
Shirling and Polk families
    see Polk and Shirling families, 174
Shirling, Milo Rex, Jr., 175
Shirling, Virginia Dallis "Ginger" Stallings
    see Stallings, Virginia Dallis "Ginger," 175
Short Farm, 35–38, 188
Short, Barbara, 36–37
Short, Benjamin F., 36
Short, Benjamin Franklin, 188
Short, Bonnie Larue, 36–37
Short, James Cotton, 36–37
Short, Jesse Armistead, 188
Short, Jesse E., Jr., 36
Short, Jesse E., Sr., 36
Short, Jesse Edlin, 188
Short, Joan, 36–37
Short, Lucile, 36
Short, Tennie Boyd, 36
Short, Virginia, 36–37
Slave cabin, 20, 23
slave dwelling, log, 179
Slave life, 48
slave, born, 187
Slavery, 47–48
Slaves, 36, 154
Slaves, freed before Civil War, 184
Smith Brothers Farm, 5, 102–103
Smith-Lever Act of 1914, 136
Smith, Annie Hazelwood
    see also Hazelwood, Annie, 102–103
Smith, Frank Erwin, 102
Smith, Fred Riley, 102–103
Smith, John D., 102
Smith, Kathleen
    see Ogilvie, Kathleen Smith, 99–100
Smith, Madelyn, 13
Smith, Sallie P. Cathey
    see also Cathey, Sallie P., 102
Smith, Thomas P., 102–103
Smith, William Franklin, 102–103
Smithson-McCall Farm, 74
Smithson, Alice
    see McCall, Alice Smithson, 74
Smithson, C.M., 152
Smithson, Charles E., 74
Smithson, Charles T., 74
Smithson, Clement, 74
Smithson, Jane Giles
    see Giles, Jane, 74
Smithson, Martha, 74
Smithson, Mattie Jane

see Luster, Mattie Jane Smithson, 155
South Harpeth & Franklin Turnpike, 36
Spanish Flu epidemic, 134
Spanish-American War, 66
Spann family, 112
Sparkman, Alice Jones
 see also Jones, Alice, 107–108
Sparkman, Joe, 107–108
Sparkman, Ollie Jo, 107–108
Speaker of the Senate, 148
Lieutenant Governor, 148
Spring Hill, 129
Spring Hill High School, 130
springhouse, 21
Sprott and Grigsby families
 see Grigsby and Sprott families, 180
Sprott, Emma Catherine
 see Bond, Emma Catherine Sprott, 123–124
Stallings, Leslie Osburn
 see Osburn, Leslie, 175
Stallings, Virgil, 175
Stallings, Virginia Dallis "Ginger"
 see Shirling, Virginia Dallis "Ginger" Stallings, 175
Starnes, Vinnie
 see Hatcher, Vinnie Starnes, 146
State Route 100, 104
Statistics, agriculture, 2, 89, 135–136
Steele, Cora
 see Bond, Cora Steele, 84, 115
Steele, Elizabeth, 83, 85
Steele, Mary Elizabeth, 84–85
Steele, Moses, 83
Steele, Susan Moore
 see also Moore, Susan, 83
Steele, William A., 83, 85
Steele, William Alexander, 83
Steele, William Alexander, Jr., 84
Stephens, Pauline
 see Gillespie, Pauline Stephens, 11
Stewart, Carolyn, 44
Stewart, Marcus, 44
Stewart, Mark, 44
Stewart, Scott, 44
Still House Hollow, 36
Stokes, Hilda
 see Barker, Hilda Stokes, 121
Sugar Ridge, 130, 187
Sullivan Family, 149
Sullivan Farm, 104–105
Sullivan-Givens Farm, 149–151
Sullivan, Allen Judson, 104
Sullivan, Andrew Jackson, 104
Sullivan, Ida Dora Pewitt
 see Pewitt, Ida Dora, 104

Sullivan, Imogene "Jean"
 see Bledsoe, Imogene "Jean" Sullivan, 158–159
Sullivan, J.T., 143
Sullivan, James Carol, 165–166
Sullivan, Jeffrey, 105
Sullivan, John Ensley, 105
Sullivan, John Ensley "J.E.," 104
Sullivan, Lenar Florence Deal
 see also Deal, Lenar Florence, 105
Sullivan, Lillie A. Tidwell
see Tidwell, Lillie A., 165
Sullivan, Linda Tucker
see Tucker, Linda, 105
Sullivan, Lucinda
 see Givens, Lucinda Sullivan, 190
Sullivan, Matilda Jane Tidwell
 see Tidwell, Matilda Jane, 149–150
Sullivan, Nancy Jane
 see Fisher, Nancy Jane Sullivan
 children of, 166
Sullivan, Nelle Edna Walker
 see Walker, Nelle Edna, 105
Sullivan, Ora
 see Givens, Ora Sullivan, 150
Sullivan, Owen Thomas "Tee," 149–150
Sullivan, Sandy, 105
Sullivan, Vickie, 105
Sullivan, William "Houston," 105
Sullivan, William Earl, 105
Sullivan, Zachariah Joseph, 158
Sycamore community, 51

T

tenants and sharecroppers, Black, 154
Tennessee, formation, 7
Tennessee Agricultural and Industrial College, 136
Tennessee Agricultural Enhancement Program (TAEP), 56, 85
Tennessee Agricultural Museum, 140
Tennessee Cattlemen's Association (TCA), 56, 130
Tennessee Commissioner of Agriculture, 68, 71
Tennessee Department of Agriculture, 140
Century Farms Program, 176
Tennessee Department of Agriculture (TDAG), 3
Tennessee Farm Bureau, 130
Tennessee Green Belt, 85
Tennessee House of Representatives, 140, 148, 178
Tennessee State Library and Archives, 145
Tennessee State Normal School, 98
Tennessee State University, 136
Tennessee Walking Horse, 99, 101
Theta community, 51
Thomas, Robin Carol McCanless
 see also McCanless, Robin Carol, 96

Thomas, Todd, 96
Thompson Station Methodist Church, 184
Thompson Station Post Office, 121
Thompson, Sandra
    see Glenn, Sandra Thompson, 12–13
Thompson's Station, 178
Thornton, Jimmy, 168–169
Tidwell, Lillie A.
    see Sullivan, Lillie A. Tidwell, 165
Tidwell, Matilda Jane
    see Sullivan, Matilda Jane Tidwell, 149–150
Tidwell, Sarah Jane
    see Sullivan, Sarah Jane Tidwell, 104
Trevecca, 98
Triangle School, 142–143
Trice, Blythe, 117
Trice, Catherine Miranda, 117–118
Trice, Charlie, 117
Trice, Grover, 117
Trice, J.G., 117
Trice, Mary Louise, 117
    see Hargrove, Mary Louise Trice, 117–118
Trice, Sam, 117
Trice, W.J., 117–118
Trinity community, 144
Triune, 95–96, 112, 153
Triune/Nolensville, 113
Tucker, Linda
    see Sullivan, Linda Tucker, 105
Turner, Mary Jane
    see McFarlin, Mary Jane Turner, 161

U

U.S. Army, 130
U.S. Marine Corps, 159
Union Valley Home Demonstration Club, 143
United States Colored Troops, 87
United States House of Representatives, 66
University of Tennessee, 70, 99, 130, 136
USDA, Farm Service Agency, 175
UT Extension, 56

V

Valley View Farm, 62–64
Van West, Carroll, Dr., 3
Vanderbilt, 142
Vanderbilt School of Nursing, 100
Vanderbilt University, 69
Vantrease, Margaret
    see Bond, Margaret Vantrease, 84
Vaughn, Kay Pewitt
    see Pewitt, Kay, 132–133

W

Waddell Hollow, 168
Waggoner, Joyce M.
    see Jefferson, Joyce M. Waggoner, 82
Walker Farm, 142–143
Walker, Carl "Bubba" III, 143
Walker, Carl H., Sr., 142–144
Walker, Dewar, 142
Walker, Era, 142
Walker, Era Frances Overbey
    see Overbey, Era Frances, 143
Walker, Frank D., 143
Walker, Geneva, 142
Walker, George "W.G.," 143
Walker, Harriet Beech
    see also Beech, Harriet, 143
Walker, Nancy Peck
    see also Peck, Nancy, 142–143
Walker, Nelle Edna
    see Sullivan, Nelle Edna Walker, 105
Walker, Virgil, 142
Walker, W.G., 142
Walker, William Thomas, 143
Wallace, Allean Harper
    see also Harper, Allean, 152–153
Wallace, C.T., 152
Wallace, Ennis Core, Sr., 152–153
Wallace, Ennis, Jr., 153
Wallace, Kenneth, 153
Ward Seminary, 17
Ward-Belmont College, 17
Ward, Eliza Hudson, 17
Ward, Viola
    see Sanford, Viola Ward, 63
Ward, William E., 17
Washington, George, 6
Watkins Institute, 142
Watson, Alice
    see Carter, Alice Watson, 187
Webb, Rachel
    see Ogilvie, Rachel Webb, 21
Welch, Wanda Bledsoe
    see Bledsoe, Wanda, 159
West End School, 128
West Harpeth River, 35
Westbrook, 188
Western Electric, 110
Western Highland Rim, 104
Wheeler, Joe, 66
White, Cindy
    see Gentry, Cindy White, 76, 78–79
White, Era Ann Sanford
    see also, Sanford, Era Ann, 63
Whitley, Talitha
    see Williams, Talitha Whitley, 25
William Steele Farm, 83–86, 115
Williams family, 120
Williams, Cynthia German
    see also German, Cynthia, 25

Williams, Henry Mortimer, 25
Williams, Janice, 26, 28
Williams, John Joseph "Joey," 26
Williams, John, Jr., 26–28
Williams, John, Sr., 25–28
Williams, Mary Armenia Gillespie
    see Gillespie, Mary Armenia, 25
Williams, MaryLynn, 25–28
Williams, Oscar Fitzallen, Jr., 25
Williams, Sarah, 26, 28
Williams, Stacey
    see Givens, Stacey Williams, 150–151
Williams, Talitha Whitley
    see Whitley, Talitha, 25
Williamsburg, 120
Williamson County
Census statistics, 1800 and 1810, 7
Williamson County Cattlemen's Association, 103
Williamson County Cavalry, 16
Williamson County Commissioner, 29, 30, 31, 110, 128
  Williamson County Court, 145
  Williamson County DHIA Board, 140
  Williamson County Election Commission, 110–111
  Williamson County Fair, 140
  Williamson County Farm Bureau, 103
  Williamson County Farm Service Committee, 140
  Williamson County Livestock Association, 140
  Williamson County School Board, 140
Willow Creek Farm, 168–171
Willow Run Farm, 144–145
Wilson Family Farm, 123–125
Wilson, Aaron, 123

Wilson, Aaron C., 124–125
Wilson, Emma Ida
    see Jordan, Emma Ida, 144
Wilson, Gladys Bond
    see also Bond, Gladys, 123–125
Wilson, Landon, 124–125
Wilson, Lucas, 124–125
Wilson, Lynn, 123
Wilson, Lynn Chester
    see also Chester, Lynn, 124–125
Wilson, Riley, 124–125
Wilson, Samantha, 124–125
Women's Army Corps (WAC), 175
Wood, Allen F., 62
Wood, Mary Ann
    see Sanford, Mary Ann Wood, 62–63
Wood, Sarah, 62
Woodland Farm, 183
Woodside, Lynn Ogilvie
    see Ogilvie, Lynn, 100
Woodside, Mike, 100
World War I, 21, 30, 66, 70, 84, 123–124, 134, 147
World War II, 25, 70, 92, 121, 135, 145, 159, 169, 175
WSM Tower, 139, 141

Y
Yoest, Charlee, 72
Yoest, Chuck, 71
Yoest, Hatcher, 71–72
York, Maude
    see Green, Maude York, 39, 41

About the Publication Team

Caneta Skelley Hankins, Marcia Fraser, Elaine Warwick, and Rick Warwick at the Sage Awards, November 7, 2025, in Franklin.

Caneta Skelley Hankins

Caneta Skelley Hankins, a native Tennessean and eighth-generation resident of Williamson County, holds degrees in English, History, and Historic Preservation. When the Center for Historic Preservation at Middle Tennessee State University was established in 1984, she served as the Projects Coordinator and then as Assistant Director from 2001 until her retirement in 2013. Among other responsibilities, Hankins served as Director of the Tennessee Century Farms Program from 2001 to 2013. Another particular interest is Scots-Irish resources in Tennessee. This led to consultancies and cooperative efforts over two decades with the Department of Education in the Republic of Ireland and the Museums of Northern Ireland (UK).

Among her numerous publications is *Barns of Tennessee*, co-authored with Center colleague Michael Gavin, in 2008. They also collaborated on *Plowshares and Swords: Tennessee Farm Families Tell Civil War Stories*, published in 2014. After retiring from

MTSU, her research and publications have focused on Williamson County topics. She collaborated with Rick Warwick on *At Home with Working Folks in Williamson County* in 2018 and was the co-author with Warwick for *Barns of Williamson County* in 2019. Her article "Travelers Rest of Williamson County: World Renowned Arabian Horse Farm" was included in the *Williamson County Historical Society Journal* in 2022.

Caneta has actively served on the boards of the Tennessee Agricultural Museum and the Williamson County Fair, where she coordinated the annual Century Farms exhibit and dinner for 15 years. A church musician for over fifty years, Hankins is the organist at Hillsboro United Methodist Church in Leiper's Fork. Beyond these interests, you may ask about her dogs and women's basketball.

Rick Warwick

Rick Warwick, president of the Williamson County Historical Society, was appointed Williamson County Historian in 2017 by the County Commission. Known by all simply as "Rick," he was honored in November of 2025 with the Sage Award from the AgeWell organization of Middle Tennessee, recognizing him as a leader who has made a lasting impact on his community and Tennessee. The award noted Warwick's "dedication to improving the quality of life in Williamson County through education, volunteer service, and historic preservation."

Rick is the well-known author of a myriad of local history books and is the keeper and collector of thousands of historic images of people, places, and events that tell the stories of those who have contributed to the county's rich past. During his tenure as editor of the *Journal* of the Williamson County Historical Society, many topics were explored, and Rick encouraged and assisted many contributors while also authoring articles and publishing books each year. Rick's diligent and impressive body of work and knowledge is recognized and appreciated in Williamson County and across the state and beyond through his collaborations with individuals and organizations, his presentations at many conferences, and his service on the board of the Tennessee Historical Commission.

Marcia Fraser

Marcia Fraser, a native of Georgia, was well prepared for a career in library and information science through her graduate work in Library Science at Dominican University in River Forest, Illinois. In her most recent position, she served as Special Collections Librarian at Williamson County Public Library. In this role, she helped local authors format and publish their works after the library invested in a professional publishing platform for its patrons. During her tenure at WCPL, she compiled and published a local history book, *Excellent Citizens and Notable Partings*, in 2020. Marcia, now retired, has found this work particularly rewarding and continues to offer help to the Williamson County Historical Society in editing, layout design, and publishing local history books. She is grateful for the opportunity to give back to her beloved adopted community, which she has been a part of for 45 years.

www.ingramcontent.com/pod-product-compliance
Lightning Source LLC
Chambersburg PA
CBHW051328110526
44582CB00003B/87